Youth Prolonged
Old Age Postponed

Robert Weale

King's College London, UK

Youth Prolonged
Old Age Postponed

Imperial College Press

ICP

Published by

Imperial College Press
57 Shelton Street
Covent Garden
London WC2H 9HE

Distributed by

World Scientific Publishing Co. Pte. Ltd.
5 Toh Tuck Link, Singapore 596224
USA office: 27 Warren Street, Suite 401-402, Hackensack, NJ 07601
UK office: 57 Shelton Street, Covent Garden, London WC2H 9HE

British Library Cataloguing-in-Publication Data
A catalogue record for this book is available from the British Library.

YOUTH PROLONGED: OLD AGE POSTPONED

ISBN-13 978-1-84816-507-6 (pbk)
ISBN-10 1-84816-507-2 (pbk)

Typeset by Stallion Press
Email: enquiries@stallionpress.com

Printed in Singapore.

Foreword

Robert Weale addresses many important issues in a refreshing and attractive way. He points out that serious and difficult subjects, such as the increased likelihood of age related, and often seriously debilitating, diseases are the downside to the triumph of increased longevity, which is now something that most of us can hope for. He demonstrates that the new approach can give young people knowledge about growing old and counter the ignorance and negative attitudes towards old age that are still widely held. A better understanding between old and young people could go a long way, not merely to improving life for older people but, as importantly, ensuring that young people, in turn, would reap greater benefit from their own old age. This should be required reading for young people, whose own future and experience of ageing might well be dramatically improved as a result.

Baroness Sally Greengross (Former Director-General of Age Concern England), 2009

Contents

Foreword   v

Introduction   xi

Chapter 1   **What's the Problem?**   1

Chapter 2   **Language as a Barrier**   5
             'I can't hear you'   7

Chapter 3   **Dress and Appearance**   9
             Disguising the years   10
             Failure?   13

Chapter 4   **Digging Up the Past; or, Where**   15
             Do We Come From?
             Were our ancestors healthier than we are?   16

Chapter 5   **Ageing Factors**   17
             More about appearance: Tell-tale   20
             signs — Sun, smiles and smoking

Chapter 6   **Guessing by Experts**   24

Chapter 7   More About the Skin, Posture and Bones   27
The skin   27
Posture and muscles   28
Bones, and (almost) all fall down   30

**Chapter 8   Biomarkers; or, The Countdown to the   32
End — Men and Women, Life-Expectancy**

Chapter 9   The Eyes Have It   35
Focusing   36
Colour change   38
The eyes can show their age   41

Chapter 10   Thought for Food   44

**Chapter 11   Why Do We Age? Is It a Matter of   49
Biological Economics?**
Biological economics   50

Chapter 12   Elements   54
The keys to ageing?   57

Chapter 13   Some Age-Related Diseases: Risk Factors   58
Alzheimer's disease   59
Common problems of the bones   63
Risk factors for some cancers   68
    Bone cancer   70
    Brain cancer   70
    Breast cancer and ovarian cancer   71
    Colorectal cancer   73
    Lung cancer   75
    Cancer of the prostate   75
    Stomach cancer   76

The heart at risk 78
Parkinson's disease ('the shakes') 81
Stroke: the brain at risk 82

Chapter 14 The End of Ageing **86**

Chapter 15 What Can We Do About All This? **89**
Smoking and the eyes 91
Can mental attitude play a role? 93
What else can be done? 94

Chapter 16 Summary of Chapters 1–15 **98**
Ageing (leading to some detailed reading) 98

Chapter 17 Old Age **100**

Chapter 18 Biomarkers **103**

Chapter 19 The Menopause **107**

Chapter 20 Age in the Distant Past **110**

Chapter 21 How Does Human Ageing Fit into **112**
the Animal Scheme?

Chapter 22 From End to Start **118**

References **128**

Index **131**

Introduction

"Grow old with me, the best is yet to be" — so pleaded the Victorian poet Robert Browning in his mile-long poem "Rabbi ben Ezra". Little could he have imagined that, some one and a half centuries later, the former 101st Archbishop of Canterbury, the Most Reverend Donald Coggan, would be overheard at a dinner to proclaim, with a silver-haired voice, how excited he was at the thought of death.

While this is not a universally echoed opinion, ageing, as distinct from dying, has become fashionable because many more people partake of the experience than was true only 100 years ago. This is due largely to medical advances, but also to the recognition that the rate of ageing has become partly controllable, at least amongst relatively affluent populations. Life-expectancy has increased faster than health-expectancy, which is why Browning's hope may take some time to reach fulfilment.

The object of this slim volume is to assist mutual introductions between those occupying opposite poles of life. It is intended for the pre-elderly from some 20 years old or under, for grandparents who might like to discuss suitable sections with the younger generations (but see Section 1), and also for intermediate age-groups. The generation gap, such as it is, is fuelled by grandparents having forgotten that they were rebels when young, and by youngsters being unable to imagine that they will ever grow old. Otherwise they would not allow older ladies to offer their seats on the bus or in the underground to wobbly old men.

Much of the world is still ageist, witness the costly fixture of retirement ages in many countries. In the United States of America,

Youth Prolonged: Old Age Postponed

this particular voodoo was broken when a Californian academic took his employers to court to fight enforced retirement — and won. The United Kingdom legislature has passed anti-ageist legislation, which has merely tinkered with the problem. It is likely, however, that, if life-expectancy keeps increasing for a while, the distribution of political power will make itself felt in favour of anti-ageism.

The burden of the text is that, while the ageing process as such is at present unstoppable, it may be possible to reduce its individual rate of progress by sheer willpower. The downside of this message is that it is unlikely to be of any value if we start modifying our lifestyle only when we retire or when we are 60 or 65, or however many years old. This is why the pre-elderly of 20 years of age or more are being addressed: by these ages the awareness of the need for an investment to be made in one's future should be apparent. While, therefore, early awareness is desirable, the early middle age can still bestow benefits. The first part of the text is intended for those who seek to know, whereas the second one from Chapter Sixteen onwards — a mere appendix — may interest those who also wish to delve more deeply into the problem. But it is hoped that both parts may prove to be of interest.

Good luck to all of us.

The following have helped me a great deal, and I am much indebted to them: Ms S. Haynes, Dr G. Hipkins, Professor W. Hodos, Dr K. Lowton, Dr J. Preston, Professor H. Ripps, Mr J. Robinson and Mr L. Sucharov. Mr. T. Weale drew Figures 1.1, 1.2, 1.3, 15.2 and 22.5.

I would also like to thank the original publishers for kindly allowing me to reproduce in this book some of their figures. In some cases, it was not possible to contact the editors, and I hope they will accept my sincere apologies. Should contact be established, due acknowledgment will, of course, be made in any future edition, or reprint, of this book.

R. W., 2009

Chapter One

What's the Problem?

Fig. 1.1. "We choose friends who are roughly as old as we are".

Throughout our lives, and particularly when we are young, we automatically choose friends who are roughly as old as we are (Fig. 1.1). The reasons are obvious. Youngsters younger than ourselves are likely to be smaller, and will play with mates who do not run any faster than they do, or climb higher than they can, or do not understand the words that we have picked up. Similarly, we are going to avoid grown-ups, who run faster than we can, climb higher than we do, and talk about things we haven't come across or understand (though we would probably be able to hold our own when it comes to discussing soccer or a particular celeb's

Fig. 1.2.   "because they are part of our childhood".

hairstyle). The reverse is also true: many a mature irate customer may have flown at a bland receptionist who connected him with a "whippersnapper" in his or her twenties when, in fact, he expected to meet a senior executive.

When, in our early years, we come across people who are "old", all this becomes much more difficult. We all know someone old when we see one, but how would we describe him or her? It is possible that the first "old" people we come across would be granny and grandpa or, more likely, these days, great-granny and great-grandpa.

But because they are part of our childhood (Fig. 1.2) they do not specially stand out — they are part of our furniture, like mother or father. It is the other old people who are the strangers, the people whose names we don't know (Fig. 1.3). It is they who are classified as being old. We may note the colour of their hair (if they have any), their slower walk, their faces probably marked by lines, maybe the odd missing tooth, and drooping shoulders. Their voices will be quieter than those of younger people. In addition they may be wearing glasses when reading, whereas in our young years the

Fig. 1.3. "'The strangers... are classified as being old".

need of them would never occur to us. Although it may be well hidden, a hearing aid may be nestling in one or the other ear or both. And sometimes the "oldsters" may be ambling along supported by a stick.

All this may be enough for us to look on them as strangers. The upshot is that we form opinions about them to make us feel comfortable, a process known as stereotyping. This word describes a way of thinking in which individuals are grouped together, not infrequently to be mocked. The grouping together serves to simplify our thinking: if we lump different people into one word-bin, we need to think only about the stereotype rather than about the differences which distinguish them.

Looking up the word "stereotype" on the Internet is a revelation: it offers quite a few surprises by what it has to say. A great deal of stereotyping occurs in connection with the elderly, which is as unjustified as when applied to other groups (or when the elderly apply stereotyping to young people). An interesting example was quoted by Sir Malcolm Rifkind, when he reviewed Menzies

Campbell's autobiography in a British Sunday paper — *The Observer* — early in 2008.

> In the last few months of his leadership [of the parliamentary LibDem party], this 66-year-old ex-Olympic athlete was portrayed in the press as a doddering geriatric with constant references to Zimmer frames, bus passes and pensions. Such is the cult of youth today that this was seen as fair game and there was little Ming [Menzies] could do about it. It was absurd when judged against the national interest. Churchill and Harold Macmillan were both in their middle sixties when they first became Prime Minister and they didn't do too badly. The US may be about to elect a 71 year old as its next President.

Not as though there were anything new under the Sun. More than 400 years earlier, Shakespeare's Earl of Gloster (who appears in *King Lear*) reads the following letter:

> This policy and reverence of age, makes the world bitter to the best of our times; keeps our fortunes from us, till our oldness cannot relish them. I begin to feel an idle and fond bondage in the oppression of aged Tyranny; who sways, not as it hath power, but as if it is suffer'd.

As implied earlier, the object of this slim book is to offer some explanation about how people become elderly, how youngsters in the long run are likely to become like them, how ageing can be delayed, and to show that, by stereotyping those with years of birth earlier than our own, we finish up by stereotyping only ourselves.

Chapter Two

Language as a Barrier

The first essential for becoming friends with people is to understand their language. On holidays abroad we may meet any number of youngsters as old as ourselves: if we don't speak their language — or they don't speak English — no contact is likely to be established between them and us. It is easy to see that this applies to other strangers, including the elderly. But this may need an explanation. It will be said quite rightly that they may speak English: why shouldn't the rule hold? The answer is that this is right, they may be speaking English, but the uses of English of the young and that of their elders are very likely to differ somewhat from each other. Languages don't stand still. New words enter them not only because of technological developments, but also because of musical compositions, songs, and inventions, because of changes in slang, simplifications — not to say erosions — in daily grammar, and so on. We may remind ourselves what effect texting has had on the form of advertisements, posters, and newspapers. If one were to have a look at 20-year old posters, one would call them old-fashioned (Fig. 2.1). But they were the last word when they were first produced, and today's perfectly "with-it" hoardings will be old hat in a few years' time.

The upshot of these and other changes is that each generation of people has its own use of language, its own vocabulary, its own way of expressing itself. This can be overridden, of course, by learning and reading, and using what the BBC calls "Received English". It used to be called "Queen's (or King's) English", and the change demonstrates what we are discussing (incidentally,

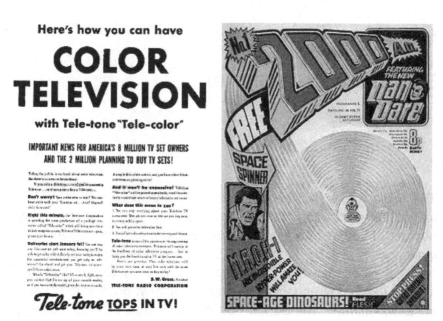

Fig. 2.1. Two printed messages: one with a year 1950 poster and the other from a 1977 publication.

people's attitude to the British Monarchy is a subtle but pervasive factor in this process of changing our language).

Just because people one might refer to as being elderly do not talk about Chavs or Goths does not make them any more peculiar to us than we are to them for using such words, which they might refer to as being "new-fangled", incidentally an out-of-date expression if ever there was one. Of course, this inter-generational barrier is easier to jump over than the one we may face on holidays abroad, since "Received English" will have been taught in our schools as likely as in those attended by older generations.

All this goes to show that the lack of understanding that has not been overcome is not one-sided, but reciprocal: it works both ways. One is, of course, perfectly entitled to view the elderly as being beyond the pale, but one has to remember that they may think the same of us (equally without reason). It is significant that, when travelling by train, the coaches all seem to us to be wobbling with the exception of one — namely the one we are in.

"You misheard me, man - I said I was bringing two ageing babes, not two asian babes."

Fig. 2.2. "They may fail to hear some or all of them" (www.CartoonStock.com).

'I can't hear you'

There may be, in addition, a more basic barrier between the young and the not so young: the latter may not only misunderstand our words, but they may also fail to hear some or all of them (Fig. 2.2).

Hearing deteriorates in an interesting manner which is shown diagrammatically in Fig. 2.3. The different lines for each age group show the loss of hearing, as measured on the left, for different sound frequencies, which are shown on the bottom horizontal. A low frequency corresponds to a low note and vice versa. dB is short for decibel, which measures relative values of power. Here, the power measured is that of the smallest sound allowing one to hear, namely the hearing threshold. The scale is not arithmetic like a taxi meter but logarithmic. This reduces the graphic impact of large values, and, conversely, magnifies that of small ones. The graph illustrates the important point that older people lose hearing of high notes at a greater rate than is true of low ones: a booming bass can be heard almost in the grave! The practical aspect of this fact is

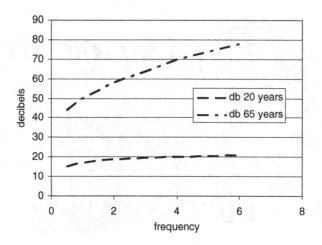

Fig. 2.3. Hearing threshold (vertical scale); frequency of test note (horizontal scale). Note that the hearing thresholds for the young vary much less with frequency than is true for older people.

that, conversely, younger people will hear very high notes well — as do also dogs — and some authorities have used this circumstance to try to disperse congregations of youngsters whom some elderly people say they are afraid of. This has led to the popular press reporting on social problems associated with Mosquito, a device which makes use of this age-related difference in hearing thresholds for different sound frequencies.

Chapter Three

Dress and Appearance

Dress offers another potential barrier between the young and those ahead of them in years. In general, fashions change all the time because this provides a means for generating work, and induces us to spend our money. But in the last 50 years or so there has been a noticeable relative change in the difference between the fashions of young and older adults respectively: they have converged towards a common ground. Some 60 years ago, after the Second World War, men used to wear hats or caps, and women also wore headgear, namely hats or scarves. In addition, women used to wear gloves (Fig. 3.1), even when pushing prams (the forerunners of today's buggies). If vain about their appearance, older people out-of-doors had the additional incentive to wear their headgear, since it covered up the absence, or grey colour, of their hair.

Even if not widowed, elderly ladies tended to wear dark clothes, as is still the case these days, for example, in Greece, Cyprus, Italy and Spain, and amongst migrants from those countries. Actually, society was in those days generally much more prescriptive: teachers were not expected to wear anything garish, and even young female employees of, for example, the Bank of England were expected to wear clothes in which the overall area of light did not exceed that of dark — a nice problem in geometry, that!

Sixty years on, it causes no surprise if, much as the younger section of the population, older men and women wear jeans; and hats are worn in clement weather only to preserve body heat, or on very formal occasions. To see a woman wearing a hat these days for no obvious reason tends to mark her as rooted in the past. No one would dream of calling her cool.

Fig. 3.1.   A photograph from the 1950's showing how men and women used to dress.[1]

The reason for these changes is not hard to find. Being 70 years old today is very different from what it was 60 years ago, not to go back as far as Victorian times. As a result of medical advances and a greatly increased understanding of what is called lifestyle (see pages 79–85), quite a few people have been able to maintain their agility, and, therefore, their physical mobility for many years in their later lives. Hence it is not surprising that they have abandoned wearing the uniforms that used to mark their parents and grandparents at the same age as elderly.

Disguising the years

The battle with the years has not, however, been confined to clothes, which, after all, mediate only part of one's appearance.

---

[1] Reproduced with kind permission from The Malhamdale Local History Group. http://www.kirkbymalham.info/KMLHG/events.html?source=malhamdale.com

Since many young women dye, or "touch up", their hair, older ones see themselves as having more reason to camouflage what is grey or white.

This has been followed during the last two or three decades by men doing likewise. Before that, dyeing one's hair was the prerogative of statesmen, given they had any, because the notion that the political realm could be mastered only with grey hair (see page 4) had turned out to be fallacious when John Kennedy, 44 years old, became the youngest President of the United States. Incidentally, he was almost twice as old as the youngest British Prime Minister, the "younger" Pitt: but in the early 1800s, fashions changed much more slowly than they do now. Also more specifically, the television screen developed — and still has — a preference for young faces. While it is difficult to erase lines from one's skin, changing the colour of hair to mimic youth is manageable, and profitable as many a supermarket has demonstrated.

Birthday greetings

This is your day though cataracts of rain
Pour down and hide the grilling birthday sun.
Keep smiling though, while lower heels
Ensure that, when — a bus to catch — you run,
You will not fall. The dye turns white hair green,
And no one offers you a vacant seat.
Don't worry, dear. Your varicose, blue vein
Is kept by Levi's trouser leg unseen.
Enjoy what's left, and never mind the brink.
This is the age: what once was up hangs down,
And starts with b: bags under the eyes,
Your buttocks, balanced by one flaccid breast,
The growing belly, pulled by gravity,
Assailed by Earth's relentless depravity.
Ignore regrets and tears and pointless sighs:
Go! Celebrate! Less food. More drink.
It's easy now. It's easy to forget
That what you've said you've said before.

Repeated repetition is a bore,
But none will tell you, set your mind at rest.
Now life expectancy is thirteen years
Below the age you say you have attained.
Forget regrets and sighs and pointless tears.
You've made it further than your cohort's age,
And, if you water them, these roses will
Outlive this day. And many of them still
To come — roses and birthdays is what I meant.

As Fig. 3.2 indicates, the process of rejuvenating one's appearance is now being pursued by a vast and far-flung industry,

"As a kid I was told, 'Act your age.'
As an adult I'm told, ' Don't look your age.' "

Fig. 3.2. The process of rejuvenating one's appearance is now being pursued by a...far-flung industry (www.CartoonStock.com).

involving not only cosmetics but also surgery of many kinds. But what the cartoon shows in addition is that, in our young years, we are also subjected to pressures, occasionally urging us to "grow up", not to "behave like a baby", or simply to follow some older role-model.

Failure?

The difference in physical — as distinct from sartorial — appearance between the elderly and our young selves is the result of changes, not the cause of them. Let us take a careful look at Fig. 3.3: it shows a young tattooed macho male on the left, tough,

Fig. 3.3. "A young tattooed macho male...shrunk, collapsing" (www.CartoonStock.com).

muscular, confident, and, on the right, 50 years later, shrunk, collapsing, barely able to stand up. To cap it all, the fashionable sun glasses of his youth have been replaced by badly fitting reading glasses. These changes are not superficial or merely skin deep: they are the result of processes which are taking place inside most of us now, which, in an important sense, started before we were born, and which, one way or other, will keep going on until the end of our lives. We all know that there are more cheerful topics to discuss than the end of one's life, but is it not possible that understanding how the differences between young and old come about might help us to see the pattern which, when we come to think of it, is a matter of inevitability?

Chapter Four

Digging Up the Past;
or, Where Do We Come From?

There was a time when ageing was unknown as we know it. No, people did not live forever, nor did they enjoy lasting youthfulness until the end of their lives. The fact of the matter is that, on average, they did not live long enough for the signs and symptoms of age to make themselves felt. What do we mean by "long enough"?

To answer that question, we have to discuss some aspects of measurements relating to the length of life. One could, of course, record the length of every life, and take the average as a measure. But the common measure is called, life-expectancy. This is the number of years attained by half the number of people in a given age group. Let us assume, for example, that, in a selected community, there were 1000 men and 1000 women born on 17 January 1915. Five hundred of them were alive on 16 January 1983 and 1988 respectively. This is taken to mean that the life-expectancy of the men at the time of their birth was 68 years, whereas for the women it was 73 years. For reasons mentioned later, these two figures are not cut in stone, having grown appreciably since nearly 100 years ago.

But, conversely, life-expectancy used to be much shorter, centuries and millennia ago. Estimates for Ancient Rome for about the first century AD are about 30 years, as obtained from inscriptions on tombstones. This estimate is probably too high for a reason not frequently mentioned, namely that only affluent survivors would have been able to afford the cost of an inscribed tombstone for their dear departed. There is a tendency for better-off individuals to live longer than underprivileged people simply because they can afford better shelter from bad weather. In addition, there may have been in Roman times an under-representation of infant deaths. Thus it is estimated that a population exposed to "natural" hazards such as poor shelter, wild animals, or environmental accidents for example would have a life-expectancy nearer the low twenties. Indeed, there is some anecdotal support for values of this order among isolated African communities, with a figure as low as 18 years having been reported.

It is clear that, if the life-expectancy of a population drops to too low a value, the families are likely to die out. A little thought will show that what is needed for survival of a population is two consecutive stretches of puberty: parents have to live long enough for their children to reach the reproductive age and to be able to do the looking-after in their turn. (The statistic "life-expectancy" will here be distinguished from an individual's "life expectation".)

Were our ancestors healthier than we are?

If millennia ago, and, more recently, in the wild, people lived on average shorter lives than they do now, it is most unlikely that they will have suffered from conditions that are liable to shorten our lives today. These are called age-related diseases: well-known examples are pneumonia, diabetes, cancer, tuberculosis, and others, generally affecting people at ages in excess of the above-mentioned life-expectancies of 20 or 30 years, which prevailed many centuries ago. At present, these diseases tend to accelerate a variety of ageing processes in later life: when lives were cut short by the raw environment in the wild these processes had not had time to become apparent. These days, there are research workers who predict that, before very long, people are going to enjoy a longevity (maximum recorded length of life) far in excess of the 120 years or so which is known to exist at present: mention has been made of 150, even 200, years, but no thought has been given to the secret hoard of yet unknown diseases that such a rise in longevity might reveal.

Chapter Five

Ageing Factors

It is important to realize that ageing does not just "happen". The processes leading to the signs illustrated in Figs. 5.1 and 5.2 are the results of at least three major influences: they are, first, traits inherited by virtue of the transmission of genes from the individual's father and mother, and secondly and thirdly, wear and tear, or environmental influences in the broadest sense. The "broadest sense" includes not just diet, environmental temperature, exposure to health hazards for example, but also antenatal influences, such as the mother's diet, the extent of her smoking, intake of alcohol, drugs, and even unspecified stress. This means that, when we wish to become parents, and will want to give our offspring the best possible antenatal treatment, we shall want to remember these potential risk-factors.

It is interesting that, early in 2008, two serious newspapers published within two days of each other reports of important, not to say grave, studies. One of them pointed out that stress during infancy, never mind if it was under-nourishment, poor parenting, or problems in the family, led to a relatively inadequate development of a child's intelligence. The other report said that poverty in early life was likely to have long-term effects on an individual's life. Since it is well known that underprivileged people have a shorter life-span than those who are better off, it is clear that the rate of ageing is a long shadow cast from early childhood.

The evolutionary aspects of ageing — the program — can evidently be modified partly by our parents and partly by ourselves. This leads to a fundamental conclusion: ageing does not start on

Fig. 5.1. 'Ageing does not just "happen"' (www.CartoonStock.com).

one's sixty-fifth or seventieth birthday but, in an important sense, even before we are born.

Unfamiliar with ideas of genetics as he inevitably must have been, the celebrated (and unhappy, if ill-mannered) Swiss philosopher Jean-Jacques Rousseau (1712–78) nonetheless wrote that we start dying the moment we are born. This showed a remarkable insight into, and understanding of, the biology of life. It revealed an understanding of how we change with the years, in a manner that does not involve sharply divided periods, but a gradually changing situation. In a sense, Rousseau was anticipated by the great

Fig. 5.2. 'The results of at least three major influences' (www.CartoonStock.com).

Roman writer and orator, Marcus Tullius Cicero (106–43 BC) who wrote what may have been the first essay on ageing. He saw it as a disease the (age of) onset of which was uncertain even though its end was very clear!

Now the genetic aspects of ageing are modelled by evolution giving rise to a genotype, i.e. the assembly of heritable or genetic attributes, to be distinguished from the phenotype's (i.e. the individual's) acquired characteristics. The latter can be very persistent, even if the individual has moved into a new environment: though transported to the Americas some 500 years ago,

African Americans have largely preserved their original biological characteristics together with important differences in a susceptibility to disease when compared with Caucasians. This goes to show that 500 years is a short time on the evolutionary time-scale.

It is, however, of considerable interest because it is now believed on the basis of good genetic evidence that mankind is likely to have originated in South Africa, in particular within reach of the sea near the Klasies River Caves along the coast of the eastern part of the Cape. Human remains with modern anatomical characteristics have been dated as 100,000 years old, i.e. some 60,000 years before the human exodus to Europe and Asia. This puts the above-mentioned 500 years that have elapsed since the African transports to the West into some sort of perspective: changes in the environment need literally thousands of years to manifest in a biological alteration.

More about appearance: Tell-tale signs — Sun, smiles and smoking

If we try and guess someone's age — and our guess should never be divulged when the someone is a woman — we have a number of clues to help us. One way or other they are superficial and linked to the skin. We have to remember that the skin is not just the outward cover of the body that may get tanned when exposed to strong sunlight, but that it also includes both hair and nails.

One of the clues to age is wrinkles and "crow's feet" round the eyes, which are the result of a loss of the protein that maintains skin elasticity. But wrinkles can be misleading. For example, the results of a study a few years ago suggested that repeated smiling ultimately causes wrinkling of the skin (Fig. 5.3). To be a little more precise, some 200,000 smiles are likely to change the skin permanently. Put into some sort of perspective, a dozen smiles a day will wrinklify the skin after 46 years! This type of ageing is, in a

"Well if they are 'laugh lines', you've got an
awful lot for such an old misery guts!"

Fig. 5.3. (www.CartoonStock.com).

sense, clearly due to wear and tear, and not primarily to the genetically induced type or due to the environment, although the loss of
elastin is also likely to play a role. As we shall see later, this is an
important protein that controls the elasticity of the skin and other
tissues.

Another potentially misleading clue leading to an erroneous
estimate of age is a skin that has been exposed a great deal to
strong sunlight (Fig. 5.4). This is less true of African people, or
their descendants, than of Caucasians, e.g. people living in or

Fig. 5.4. 'A skin that has been exposed a great deal to strong sunlight' (www.CartoonStock.com)

near the Himalayas. African skin benefits from the protection provided by a pigment in the skin, namely melanin, the same pigment that the skin of white people produces when they turn brown in the sun.

Unprotected skin exposed to strong sunlight for extended periods not only produces wrinkles, thereby creating an appearance in excess of the person's calendar years (Fig. 5.5), but it is also likely to lead to skin cancer (melanoma). This led the Australian government several years ago to mount

" HIS BIOLOGICAL CLOCK NEEDS REWINDING."

Fig. 5.5. (www.CartoonStock.com).

an effective campaign for skin protection with, one hopes, satisfactory clinical results, adequately supplemented with cosmetic ones.

Yet another environmental factor, namely heavy smoking, also tends to discolour the skin and to change its consistency in a manner that may lead it to be associated with older tissues.

Chapter Six

Guessing by Experts

" I don't mind the fact that my biological
clock is ticking..What I don't like is the
way it chimes every half hour."

Fig. 6.1.   (www.CartoonStock.com).

If the three S's, namely sun, smiles and smoking, are allowed for,
then a person's appearance can provide a good clue to age on a
statistical basis. What is meant by that? If we wish to arrive at some

Fig. 6.2. Two of many ways of measuring pressure someone can exert with his or her balled fist (rehaboutlet.com and amazon.com).

biologically useful or interesting result, it is no use confining ourselves to two or three persons, because the so-called variations or differences from one case to another are likely to obscure the target of our study. It is only by repeating the study several, or even many, times on different people that the feature common to them may rear its head out of the swamp of confounding differences. When the ages of a group of men were judged by experienced doctors by the men's appearance without their calendar age being previously made known, it turned out that those whose age was over-estimated were nearer the actual end of their lives than when the reverse was true. The over-estimates represented a biological age higher than the corresponding calendar age and vice versa.

The guessing is, however, likely to go wrong, when the immune system plays tricks, and exhibits flaws, be it for reasons of senescence (i.e. the process of ageing) or because of some clinical condition. These cannot be spotted simply by looking at someone, but require careful tests of a system that has evolved to protect us from insidious environmental hazards, namely the immune system.

Returning to the matter of biological age (Fig. 6.1), what exactly is meant by it? As a rather vague term, it is based by some workers in the field of ageing on a number of biological attributes. These

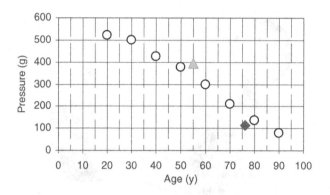

Fig. 6.3.   Hypothetical grip strength as a function age.

are liable to be chosen differently by different research workers. Suppose we measure the pressure someone can exert with his or her balled fist (Fig. 6.2). This is found to decline with age because our muscles tend to become weaker as we get older. Figure 6.3 shows hypothetical average data (white circles) illustrating the decline of grip strength (the spreads of each measurement are ignored for simplicity's sake). Now suppose two individuals are tested, *not* having been involved in the original measurements shown by the white circles. On average, the pressure of the 55-year old is about 50 g higher than would be expected (triangle) at his age, just as the 76-year old (diamond) records an average pressure lower by a comparable amount. If similar discrepancies are revealed by additional tests on the same people — and one would be looking for 15–20 reliable ones, then it might be possible to conclude that the biological age of triangle is younger by six or seven years than would be expected at his age, whereas diamond would be judged to be biologically older by some eight years than his calendar age. It is important to remember that all the measurements are assumed to be reliable, i.e. that they cannot be attributed to error or chance.

Chapter Seven

More About the Skin, Posture and Bones

The skin

But let us return to the skin. Changes in outward appearance result from physiological alterations which include both tissue structure and the underlying system of blood-vessels; these mediate the nutrition of our body organs and maintain our repair mechanisms. Exposed to the environment as it is, the skin is progressively more readily damaged, and, as mentioned above, this is partly due to the age-related loss of elastin, an important protein, which, as the name implies, underlies the elasticity of the skin. This is easily demonstrated by pinching: pinch a young skin, and it returns to its original surface hey presto! In contrast, an elderly skin is going to take its time, returning to its undisturbed surface in a manifestly lethargic manner. The lowered resistance to damage is mirrored by a slower rate of repair (when this is possible). Thus the rate of wound healing is reduced, partly owing to the aforementioned drop in the elasticity of the skin, partly owing to the loss of collagen, a ubiquitous substance that holds tissues together. The retardation in healing is illustrated in a study by Engeland, Bosch, Cacioppo, Marucha (2006) who produced precisely measurable wounds, then expressed the rate of healing at any time later by determining the ratio of wound size at that time to the original wound size. This declined rapidly to begin with, but slowed down after a week. Healing was faster for young people than older ones, with or without medication. Although there was some scatter in the data, they are statistically very significant. The authors were also able to show that, in a given time, a larger proportion of young people (aged 18 to 35 years) heal than is true of the older ones (aged 55 to 80 years). It was also found that women have larger wounds for the same-sized incision as men, and heal more slowly. This is consistent with what has been established as regards a gender difference: women have a thinner skin than men.

The repair response of the skin to inflammation is also reduced with age, so that older people may exhibit skin damage for a long time, and in addition experience, possibly indefinitely, scarring which they may have acquired late in life. A loss of capillary, i.e. very small, blood-vessels is another ageing factor, and this contributes to the relatively pallid appearance of many elderly persons who do not suffer from high blood pressure.

Risks to the skin are accentuated by its increased sensitivity to ultra-violet radiation, but this is a more general hazard: the reason is that DNA, (deoxyribonucleic acid), a nucleic acid that contains the genetic instructions followed in the development and functioning of all known living organisms, is liable to be damaged by exposure to ultra-violet radiation. Although repair-mechanisms help it to recover, this ability decreases systematically with age at the rate of just under one percent every year. The ageing skin demonstrates also another age-related feature found in other parts of the body, namely an impairment in the function of some so-called peripheral nerves. It is less sensitive to both heat and cold than is true of youthful skins. This explains why the elderly may get burned without noticing it when cooking, and, at the other extreme, suffer from hypothermia (heat loss) when they are exposed to a very cold atmosphere: a drop in body temperature from the normal of 37°C down to 30°C or less is liable to be fatal.

Posture and muscles

The perception of an individual's apparent age is also affected by his or her posture. Bent backs and hanging heads are associated with a greater age than is true of an upright posture (Figs. 7.1 and 7.2). In the absence of any disease, a relaxation of an upright

"Met my sister for lunch the other day. She's actually beginning to show her age. Of course, not her real age."

Fig. 7.1.   "Bent backs and hanging heads…" (www.CartoonStock.com).

posture is the result of the weakening of shoulder and lower muscles. In this connection, we should remind ourselves that we are descended from early primates who had not evolved the upright posture: the latter makes great demands on muscles previously not subjected to the resulting stresses. Posture also presupposes an adequate balance control (which is mediated by special bonelets in the inner ear). They all are factors which become impaired with advancing years. Age-related changes in the spine itself are also liable to contribute to postural problems; just imagine how gravitational pressure — which the spine is subjected to like the rest of the parts of the body — changed when movement on all fours gave way to the present bi-pedal (two-legged) system.

Some of these age-related changes can cause considerable pain when they involve the so-called inter-vertebral discs. The pain frequently results from a slipped disc pressing on one or more of

"Those '70s flashbacks scare the Bee-Gees-us out of me."

Fig. 7.2. "Weakening of shoulder and lower muscles" (www.CartoonStock.com).

the many nerve trunks which run down the canals of the vertebrae. The discs are made of cartilage, which is softer than the bone of the vertebrae, and discs act as shock absorbers between the spinal building blocks. The wear and tear side of ageing, however, tends to put them under such pressure that, ultimately, they become permanently deformed. They are as it were squeezed out of their original volume, a condition referred to as "slipped discs", with consequences for the maintenance of the mechanical balance of the spine in particular and body posture in general that it is easy to imagine. This, again, is a result of our having evolved from being quadripeds to becoming bipeds.

Bones, and (almost) all fall down

It is evident that posture also depends on the bones of the body fulfilling their proper function. It often comes as a surprise to learn that human bones consist of water to the extent of some seventy percent, i.e. more than two-thirds of their mass is made up of a mechanically unsupportive material. The strength of bones resides in their external shell — made of collagen (i.e. glue) and calcium. The insides of several types of bone are at least partly porous; this enables their mass to be reduced without their volume being affected. The porosity increases with age, so that the above-mentioned problems of balance are compounded by an increased fragility of the bones: when a fall occurs in later years it is liable to lead to fractures. Fractures due to falls may involve the arms when the falling person extended one of them e.g. as a result of a reflex for protection. More seriously, fractures are liable to damage one or the other of the hip bones. Causes of falls have been studied extensively, and range from defective eyesight to domestic obstacles, such as floor mats or even shabby slippers. Sometimes a merely temporary loss of balance may be the culprit.

When the loss of bone material exceeds its normal value, the abnormality is referred to as osteoporosis (see page 64–65); it is a fairly common condition particularly in post-menopausal women, and in obese persons. It is treatable, but clearly as regards the latter, prevention is better than cure.

Chapter Eight

Biomarkers; or, The Countdown to the End — Men and Women, Life-Expectancy

The ageing of muscles is associated with an interesting special feature, but let us return to biomarkers before discussing it. We have seen that biomarkers are biological attributes that change significantly with age. Grip strength (Fig. 6.3) is a well-known one, even though it is a little complicated. For example, it has been found that the hypothetical points shown in this figure start at a higher value in rural areas than in urban ones, and that they are also higher for men than for women (see below). Another biomarker is linked to breathing, and is called "vital capacity". It, too, follows approximately the trend shown in Fig. 6.3. On the other hand, some biomarkers rise with age: for example, the hardening of the blood vessels — arteriosclerosis — systematically increases with the years. Admittedly, in this case we may be dealing with an environmental cause, perhaps due to diet, but this does not affect the argument.

From a certain theoretical point of view, biomarkers showing a decline are at present more interesting than those with a rise. Many of them, perhaps the majority, decline every year at the rate of about one percent or less of their initial, young, value. This leads to the question of when they reach zero, i.e. at what age does the underlying function theoretically cease to operate on average?

Before we answer this, we have to remind ourselves that our bodies consist of cells. This is as true of the skin as of the brain.

But tissues differ accordingly as the cells may or may not remain permanent (even though capable of renewal).

In the absence of any pathology, parts of the brain age as slowly as though they could last 300 to 400 years. This is much longer than our longevity, i.e. the longest length of life, achievable at present. A possible explanation of this is that memory would not have developed in a constantly changing cerebral environment.

However, biomarkers associated with functions of tissues with dividing (somatic) cells reach their putative zero level much earlier, namely on average between 120 and 130 years of life (remember that one function of cell division is linked to repair). For a number of reasons there exists a considerable scatter in these values; some biomarkers may drop to zero, 20 or 30 years earlier, others later, by the same number of years. Moreover, authors of data obtained many years ago, but not repeated since, may not have made it clear whether or not the participants tested were smokers or not; or what their diet was; or whether there may have been controllable genetic influences; and so on. The above figures, showing an accumulation of biomarkers dropping to zero at around the age of 120–130 years, are nonetheless interesting because the longest authentically recorded life — that of Madame Jeanne Calment — exceeded 122 years. We may mention parenthetically that one of the three longevities mentioned in the *Bible* (Genesis 6:3) is 120 years, a figure, however, unlikely to be any more significant than Methuselah's 969 years; the third, being the best known, namely three score years and ten.

One of the problems with an assembly of biomarkers is that not all researchers have obtained separate results for men and women. This is surprising since the female physiology differs from that of the male. But a number of studies did record data by gender. On the face of them, the rates of decline of biomarkers are similar for the two genders. But, if they are sorted so as to form two groups, accordingly as the biomarkers relate to functions depending on muscular activity or not, there emerges a surprising result. When only muscles are involved, the putative drop-out ages of the relevant biomarkers are significantly longer for men than for women; this is to say that men's muscles age more slowly than do those of women. Conversely, when muscles are not involved, the biomarkers appear to last longer for women than they do for men. In other words, men's non-muscular biology appears to age more rapidly than that of women, although there are exceptions to this conclusion: reproductive organs provide a good example.

We have seen earlier that our genetic make-up needs to be understood in terms of our earlier evolutionary experiences. From this point of view, the difference between the genders as regards musculature is not surprising: men used to be the hunters and gatherers, and their lifestyle demanded a greater athleticism as compared with that of the more stationary women. Conversely, the difficult and demanding process of reproduction would benefit from general stamina, which is why, to this day, women are said to be the "biologically stronger" sex.

Chapter Nine

The Eyes Have It

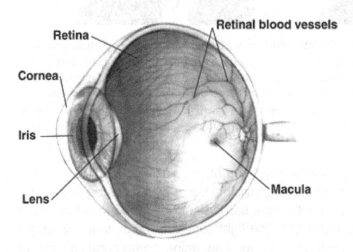

Fig. 9.1. Details of ocular structure. Visual responses generated by light in the retina (see Fig. 9.2) leave the eye via the optic nerve, shown as a tube on the right (from Wikipedia http://en.wikipedia.org/wiki/File:Human_eye_cross-sectional_view_grayscale.png).

So far, our discussion of ageing has rested here in general on people's appearance, even though hearing problems were hinted at. The tools whereby the assessment of age is largely mediated are, of course, our eyes, and, ironically, they seem to age faster than other parts of our anatomy. While a simple, general text, such as the present one, cannot go into close detail, eyes are sufficiently special to deserve a little attention on their own (Fig. 9.1).The eye has often been compared with a camera. Like the latter, it is provided with a

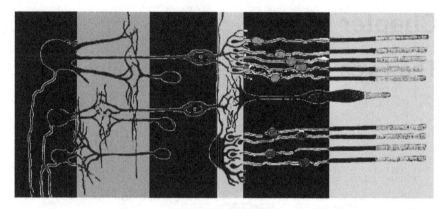

Fig. 9.2. A section through the human retina. In the eye, light enters from the left until it reaches the rods and cones on the right. The Figure shows nine rods and a single cone. The Figure also indicates the complicated network (hence 'retina') through which the neural message generated in the rods and cones has to pass to leave the eye at the blind-spot (Fig. 16.1) in order to reach the brain (from Wikipedia http://upload.wikimedia.org/wikipedia/en/2/21/Fig_retine.png).

focusing system, made up of the glass-like outside cornea and the lens, the latter being invisible because it hides behind the pupil (see Fig. 9.1). Together the cornea and the lens project an image of the outside world onto the light-sensitive retina, which lines the inside of the back of the eye. The retina corresponds to the now old-fashioned film. The cornea and lens are shown in Fig. 9.1, which is a diagrammatic slanted section through the right eye. The ocular images are upside down, but early experience has taught us that things *seen* high are *felt* to be low down so the brain settles for harmony between the two sensory impressions.

The retina is sensitive to light except at the blind-spot, which is devoid of any photoreceptor (Fig. 9.1). Because of their appearance, the receptors are called rods and cones respectively, as shown on the right of Fig. 9.2. We shall return to them on page 39.

Focusing

It is the focusing system that is one of the earliest parts of the body to signal noticeable ageing, leading to the need for reading glasses. Figure 9.1 helps to explain why. Normal eyes project onto the retina sharp images of distant objects. When the young eye looks at a nearby object, such as a book at a distance of some 30 cm, the image would be out of focus with a lens geared to a distant object. The image would be focused as it were behind the retina. However, when the retina receives an out-of-focus image, a rapid reflex changes the shape of the lens: it becomes thicker: we say that it has increased in power. The increased lenticular bulging ensures that the nearby object is accurately focused on the retina, and so enables perfect vision to be achieved.

This process is called accommodation. The ability to change the focus of the lens is developed to the highest degree in early childhood, and then declines progressively until the age of about 50 years. The declining amplitude of accommodation from childhood provides a good example of different organs ageing differently. It also explains why most people in their late forties require reading glasses for close work.

But this is only a half-truth; it applies to people living in temperate zones. Those living in hot climates are liable to need reading glasses quite a few years earlier probably because they live in an environment with a higher ambient temperature. In case you think that we are the type of animal that keeps its body temperature constant, you may like to be reminded that this is true only for the core of the body, and not its superficial tissues: the lens is part of the latter.

Thus, for example, people living near the equator are probably in need of reading glasses in their middle, or even early, thirties. It stands to reason that, if, as we are told, climate change is going to lead to a rise in temperature, the demand for early reading glasses is likely to increase. Incidentally, the great migrations of the twentieth century have revealed an interesting point: first-generation immigrants from the Indian subcontinent to the United Kingdom would be amongst those needing reading glasses when relatively young because they used to live in a hot climate. But the second and third generations have become acclimatized: their age of needing reading glasses progressively increases, approximating to that of the indigenous population.

Colour change

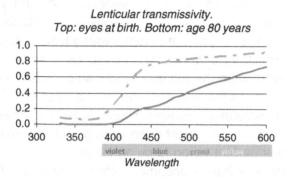

Lenticular transmissivity.
Top: eyes at birth. Bottom: age 80 years

Fig. 9.3. The lens gradually becomes yellower and yellower with advancing years.

In addition to becoming less flexible, and thus losing accommodation, the lens features another ageing sign: having started at birth with being almost colourless, it gradually becomes yellower and yellower with advancing years. This is illustrated in Fig. 9.3; transmissivity is defined as a fraction, namely the ratio of the light intensity transmitted by a transparent substance divided by the light intensity incident on it. Note the low values for the short-wavelength (i.e. violet and blue) radiations; hence the yellowing.

To be really informative, values for the transmissivity have to be obtained for the different parts of the visible spectrum, some of which are indicated along the abscissa (the horizontal axis) in Fig. 9.3; the fraction of transmissivity is shown on the left along the ordinate (the vertical axis). The lenses of babies are largely completely transparent except in the violet part and at shorter wavelengths, i.e. in the ultraviolet part of the spectrum. In contrast, the transmissivity of an 80-year old lens compares with that of a young one only in the yellow (and orange and red), but shows a marked absorption of light in the rest of the spectrum. It is easy to imagine that the violet and blue parts of the spectrum may well be effectively blocked by the time the lens has reached the potentially maximum age of approximately 120 years.

In everyday life, this age-related loss is not noticeable because of the slowness of the change. In practice, the colour change of the lens is liable to affect older people's perception of colour. The lens appears yellow because it absorbs its complementary colours, such as blue and violet; consequently such rays cannot reach the retina in sufficient strength for adequate stimulation. The result is that the perception of these colours is considerably reduced. For example, an older woman in an art gallery facing a portrait of a female sitter, wearing a blue skirt, was heard to comment on the black skirt.

There is some evidence to suggest that this loss in perceiving blue and violet has had some effect amongst painters. There is a tendency for them to use all the colours of the rainbow when they are young. But, as shown by the paintings which the Italian Titian and the Dutch Rembrandt produced in their late years, the artists seem to prefer "warm" tones with a predominance of reds and browns. A similar tendency to use warm tones later in life has been demonstrated by spectrometric measurement in connection with the 20th century French artist Georges Rouault.

In addition to the lenticular filter, many people's eyes contain another violet and blue absorber, dispersed in the retina itself. It is called the macular pigment because it is concentrated in the cen-tre of the retina where it appears as a yellow spot (Latin: *macula lutea*). Its concentration may increase with age, but the cause is probably dietary, i.e. environmental: it is said to tend to accumulate with the ingestion of green vegetables such as spinach and broc-coli. As just mentioned, its distribution in the retina is not uniform, Figure 9.1 shows the macula where the cone concentration is at its highest and where rods are altogether absent. The macular centre of the retina is the area of sharpest vision and acutest colour vision.

It can be tentatively concluded from this that the macular pig-ment probably exerts a protective function largely of the cones (the other type of photo-receptor, namely the rods are active only at low levels of illumination, and appear to be less at risk.) The accumu-lation of the macular pigment is, however, a double-edged sword: while the increase may enhance the protection of the central part of the retina, the fact that the pigment progressively absorbs more and more violet and blue light leads to a deprivation of the cones of a large fraction of the stimulus entering the eye. On balance, protection is more important than deprivation. The reason is that there are suggestions that the absence or depletion of the macular pigment may be associated with a sad disease, namely macular degeneration or age-related maculopathy. Its causes are still being discussed, but a possible age-related factor is that the retina accumulates waste-products due to its prolonged and repeated exposure to light. While not leading to complete blindness, the condition entails a loss of fine detail perception and of colour vision, making reading virtually impossible, and watching television rather difficult. The condition is caused by damage to the cones which eventually lose their function. Part of the problem may be self-inflicted, since smoking tends to compete with the cones for the antioxidant, i.e. protective, characteristics of the macular pigment: smokers run twice to four times the "normal" risk of coming by the above disease.

The etching

Saskia had died. Hendrickje took her place
In Rembrandt's bed and heart and eyes.
Maid though she was, she was a model model, too,
And cut into a copper plate.

The print hangs, lit, on Annie's wall
(She had discovered it when, after Daddy's death,
She had rummaged through a leather folder
Full of loose prints, and had it framed,
And loved it).

Day in, year out, she saw it, ever new, even though
It dated from long centuries ago.
The well arched back, the plump left arm,
The nearly hidden face,
In contrast with the background, dark and indistinct,
Offered a harmony that sang in Annie's heart.

Then, one day, it was faded. The dark
Had changed to grey, the lines to fuzz,
The shapes unshaped as if of Plasticene:
The magic vanished as if it had been magic.
Worse still, the signature was hard to see.
As time went on, the faded etching was not alone:
When Annie shot her gaze at her watch it
Vanished.
Her plants, just like the etching, were visible,
But only out of the corners of her eyes.

She bore fate well, helped by a Cyrano
Who visited this bruised Roxane, and told her
Of the daily horrors of the world which she
No longer saw in papers cancelled long ago.
Each day, and every day she waited for the hour
When a persistent bell would say "He's here".
Cyrano came, regularly, as if propelled
By some internal clock.

Then, one morning, Annie could not see
The etching, not even out of the corners
Of her eyes, even though the rest of the world
Was there. Was there.

That afternoon, Cyrano did not ring the bell,
Nor ever came again.

The eyes can show their age

But while the above internal age-related changes in the lens and retina have been established by careful and extensive measurements, they are not visible from the outside without special apparatus.

On the other hand, a careful comparison of teenage with older pupils of the eye shows that, as one ages, the area of the pupil becomes progressively smaller (Fig. 9.4). In point of fact, the

pupillary area (sq mm)

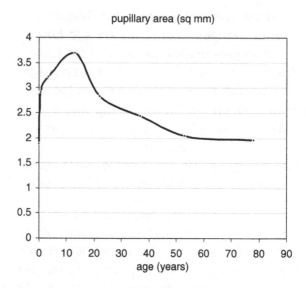

Fig. 9.4. Light-adapted pupillary area as a function of age. In the dark the pupillary area can be enlarged three or four times (after http://cnri.edu/coursedemo/ Pupils_and_Pupillary_Symptoms/Pupils_and_Pupillary_Symptoms.htm).

decline of its area would probably be visible in early childhood but for the growth of the eyeball until about the age of puberty. Figure 9.4 shows not the age effect on the pupillary diameter, as is the case with most publications, but rather that on the area. The reason is that the pupillary area, not the diameter, is proportional to the quantity of light that enters the eye, and hence is the more informative of the two quantities. Note that the area reaches a maximum at about puberty, and then starts to decline.

It is also worth mentioning that the area of the pupil is controlled by the antagonistic action of two muscles in the iris of the eye, namely the dilator and sphincter. These allow it to dilate and to constrict respectively (the pupil dilates in the dark, and constricts when the retina receives a great deal of light). The above-mentioned age-related decline in the pupillary area is the result of the dilator degenerating with age faster than is true of the sphincter. A consequence of this decline in the area of the pupil means that the quantity of light entering the eye decreases with age. During full daylight hours this is unlikely to affect how well people can see. But in utter darkness, when the smallest amount of light may be important, older people may find themselves at a disadvantage as compared with young ones, and may need special consideration. Thus it is no good for a young lighting engineer to provide illumination, for example, for a corridor in an old people's home to suit his own eyes at levels acceptable to him; old residents, with smaller pupils, not to mention much yellower lenses than his, might easily fail to spot an obstacle with potentially disastrous consequences (see Chapter Seven). It follows that, with an increasingly older population, public places need to be illuminated at night not in some average manner suitable for a hypothetical population, say, 45 years old, but for those who are at risk in the dark if lighting is insufficiently adequate for their retinae. A convoy of ships has to move at the rate of the slowest ship: otherwise stragglers will be left behind.

Chapter Ten

Thought for Food

"If you don't like it, it's okay to eat it."

Fig. 10.1.   (www.CartoonStock.com).

In order to live long enough to reach the stage when ageing becomes significant, the individual needs to take in food. It is interesting in this context that numerous experiments on a number of mammalian and other species have shown that longevity is best increased by dietary restriction (Fig. 10.1).

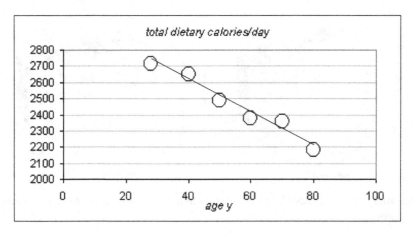

Fig. 10.2. Human calorie-restriction data.

It would obviously be unethical to carry out dietary experiments on human beings (*pace* animal lovers), but we shall see later (see pages 94–95) that there exists a community who appear to limit their food intake voluntarily, with amazing results on their longevity. Food can, of course, play a dual role in its function of maintaining life: a number of dietary factors have been shown to contribute to the appearance of some diseases. To the extent to which the latter may appear late in life, they may, indeed, curtail it. The mechanism by which dietary or caloric restriction (Fig. 10.2) appears to operate is that it delays not only maturation (in animals) but also the decline of types of immunity. The resulting period of grace appears to delay the development of neoplasms (potentially cancerous growths): this could therefore contribute to an extension of longevity. Dietary restriction appears to operate also in individual organs. Thus it has been shown that an inner layer of the cornea of the (rat) eye loses its cells with advancing age. Following dietary restriction the rate of loss is significantly slowed down. However, the actual mechanisms of action of dietary restriction are, as yet, unknown. But we have to face the paradox that, provided they avoid a noxious diet, privileged societies who can command unlimited dietary resources will far outlive under-privileged ones who may be on the verge of

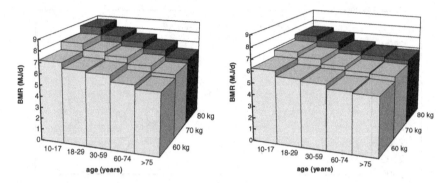

Fig. 10.3. Basal metabolic rate as a function of age and body weight. L: men; R: women.

starvation. The reason is that starvation diets are usually poor in protein, vitamins and essential nutrients, and consist of carbohydrate rich staples such as rice and porridge of various kinds. In contrast, calorie-restricted diets are specially formulated to reduce complex carbohydrates (and fats) but maintain protein, vitamin and other essential nutrients, so starvation diets are unlikely to provide potential benefits. Potential benefits due to the enforced dietary restriction of the under-privileged sections of humanity are probably far outweighed by disease and everyday environmental hazards, not to mention an inadequacy of above necessary food stuffs, all of which are under some control in the privileged parts of the world.

The basic food intake decreases with age for a number of reasons. The basal metabolic rate (BMR), the energy the body expends while at rest, declines in both men and women (Fig. 10.3).

At first sight a resting body would not be expected to lose energy. However, many of its muscles are active when the body is at rest. The heart keeps beating; muscles involved in breathing remain in action; also the body temperature has to be maintained; and the brain consumes energy simply to keep going. Again, in later years, appetite for food declines partly because the sense of smell becomes impaired, and energy demands decline because

Fig. 10.4. (www.CartoonStock.com).

bodily movements tend to slow down. As regards fluid intake — and by fluid is meant water! — the elderly do not exhibit a reduction in thirst under normal conditions (Fig. 10.4).

However, when stressed, for example by fluid deprivation or by physical exercise in a warm atmosphere, they are liable to experience a reduced sensation of thirst and hence to reduce their fluid intake. This seems to be a matter of rate rather than absolute volume since, in the end, full fluid restoration takes place. In general, an adequate fluid intake by the elderly (30 ml of water/kg bodyweight/day) is important because of the danger of dehydration. In some respects, the drinking behaviour and response to thirst represent a reversion to childhood, although not many data for children appear to exist.

The ingestion of fluids is also affected by that of minerals: sodium, contained in salt, is particularly relevant. It stimulates the sensation of thirst, and, in many people, needs controlling because an excess of salt, however tasty, is liable to raise the blood pressure in the long run with potentially undesirable results.

Chapter Eleven

Why Do We Age? Is It a Matter of Biological Economics?

This is not an easy question to answer, witness the fact that a review published almost 20 years ago listed 300 theories of ageing. Not a few more have appeared since then. During recent decades theories leaning on evolutionary ideas tended to find more support than, for example, so-called teleological concepts. Thus the notion that we age so that our globe may not become over-populated envisages an objective in ageing, which is incompatible with evolutionary principles: no mechanism, other than that of speculation, appears to have been advanced in its support. We shall see later that we have to look for an explanation of ageing at the cellular level of our bodies.

Ageing has been defined classically as the result of irreversible organic changes which make it harder for an individual to reproduce, to respond to environmental changes, to benefit from food intake, and to move. This, however, is a list of manifest characteristics which are consequences, not causes, of the ageing process. However, it is quite clear that some sort of adaptation to ageing can be learned: we abandon risky types of sport when our reaction time has slowed down or our muscles can no longer sustain juvenile exertions. In the mental field, repetition is seen as a palliative for a changed memory. And a loss of appetite may avoid overeating when the BMR (see page 46) has decreased. Thus, although, a reduced adaptability is characteristic of ageing, we are capable of learning, notwithstanding the popular view that one cannot teach old dogs new tricks.

Biological economics

If one subscribes to Darwin's theory of evolution, one will probably postulate that, though evolutionary changes are matters of chance, some of them involve responses to the environment that, by natural selection, favour the survival of a given species, in the present case, us. At the same time, evolutionary pressures are extremely mean: their economics are based on what they can get away with, while still achieving the above objective.

This is illustrated by the disposable soma theory. Somatic cells are the building bricks of the skin, hair, blood vessels, muscles, nerves etc, and are to be distinguished from germ cells which form ova (eggs) and sperms, i.e. reproductive elements. Now the disposable soma theory suggests that a species is either long-lived and produces a relatively small number of offspring, or vice versa. Thus the economic benefits are invested either in the individual, enabling it to produce resistant offspring, or in vulnerable masses of offspring, giving chance a chance, but not both. The consequence is that, if a long-lasting individual is to result, biological investment can be effective in only relatively few offspring, optimizing their chances of survival in the early postnatal period. Whereas, if the parents' lifespan and hence the reproductive period are short, limited biological resources are more expendable, and offspring can be scattered far and wide, since there is a finite probability that, at least, some of the off-spring will survive hostile conditions, predators, and disease.

On a broad canvas, predator species will qualify for membership of groups breeding small numbers, and live relatively long, whereas their booty will belong to the opposite one.

The principle of applied economics is further well illustrated by the repair mechanisms of our body (see page 33). For example, it has been found that, if DNA is exposed to ultraviolet radiation, it is damaged. However, in young persons it has the ability to recover, an ability which declines with age every year at a rate of just under one percent of its original magnitude. In other words, the repair process is adequate for maintaining the function when the body is young, but little biological investment is made when the body is past its prime. We may wonder whether this is a matter of cause and effect. It actually allows us to view ageing as a result of the failure of the body to maintain its repair processes. The latter are probably associated with the activity of stem cells. What modern medicine is doing is to back the processes up, to supplement their deficits, or to provide substitutes if necessary and if possible.

The idea that repair processes may help to control the dynamics of life leads to the question whether a glass of water is half full or half empty. If a repair process is active it may help to prolong life. For example some versions of the so-called Klotho gene control longevity, and, with it ageing, because they may have an effect on blood flow, as they are linked to stroke and heart disease. However, one is playing with words by calling them longevity genes when, in fact, they are part of the repair system of our body.

Although the idea of a single ageing factor has not been ruled out (there have been suggestions that the continued production of free oxygen in human tissues promotes ageing), the balance of opinion is strongly in favour of the view that a number, possibly a very large number, of genes are likely to be involved. If one believes that more than one factor contributes to the process of ageing, there inevitably arises the question of whether they are interdependent. This might evidently reduce to the notion of a single factor, such as the absence of caloric restriction (see page 45), and is not held to be likely. In other words, in order to qualify as determining agents, contributory factors have to be independent from one another.

A strong argument in support of a polygenic (multi-causal) theory of ageing is its variability. In statistical analyses, the latter manifests by a scatter of data that significantly increases with age. The reverse also applies. For example, like that of the rat (see page 45), the human cornea of the eye loses cells from one of its internal layers without the rate of loss being age-related; the age-related scatter of the measurements likewise tends to be similar at all ages tested.

Currently, a number of theories attempting explanations of causes of ageing are in fashion; all of them contain some truth, but none of them the whole truth. Thus during the second half of the previous century it became clear that free oxygen can be very noxious. Intuitively the gas is viewed by the layman as beneficial: haven't we learned in school that it forms one fifth of a given volume of air, and, as Lavoisier demonstrated before being guillotined by the French Revolution, that it is essential for life? Life yes, but bodily tissues are another matter. Thus the presence of *the antioxidant* super-oxide dismutase (SOD) in a tissue protects it against potential ravages by free oxygen radicals (i.e. compounds containing oxygen). Free radicals are highly reactive chemical molecules which result as by-products of all cell functions. They are liable to damage not only cell membranes, but also proteins and also DNA. They can also lead to the deposition of an "ageing pigment", namely lipofuscin, which is easily detected, for example, in the eye. When lipofuscin is illuminated with ultraviolet light, it emits a yellow-green light of its own: if such a light is directed into the eye, lo and behold! the yellow-green is brighter in older eyes than in young ones, thus providing a measure of ageing. The oxidants are kept under control by antioxidants or scavengers, but, as we age, the latter are liable to lose out to the former. The possibility has been mooted that dietary restriction (see pages 44, 45) raises anti-oxidant efficiency in combating free oxygen radicals or indeed reduces the production of free radicals in living cells: this is established as a favoured mechanism.

The noxious effect of free radicals is liable to be accentuated by smoking, toxins and the consumption of saturated-fat products, in addition to exposure to infectious diseases.

Arguably the accumulation of ageing pigments could be viewed as being linked to chronological, rather than biological, ageing, The reason is that there may be present an element of "wear and tear": for example the retinal accumulation of lipofuscin is at least partly attributable to the amount of light an eye has received.

An analogous argument applies to the ageing of teeth, which are being ground down, and, also at least in part, to the ageing of muscles which may be overworked.

Another theory rests on changes in proteins, the bricks of our body tissues. Proteins are not laid down for life, but have a life of their own, which entails renovation and reformation. The latter are based on patterns laid down by the more elementary chromosomes. Thus the changes follow at certain intervals, an increase in the delay between which has been attributed to ageing. In addition to rate changes, the chromosomal patterns may themselves alter as a result of developing errors in the course of their restructuring, another circumstance associated with ageing. "Associated" is used advisedly since cause and effect are apparently interchangeable. In the light of what was said earlier about the probably fundamental role played by repair mechanisms, it is, however, on the cards that pattern errors in the re-synthesis of chromosomes are the more likely immediate cause.

Chapter Twelve

Elements

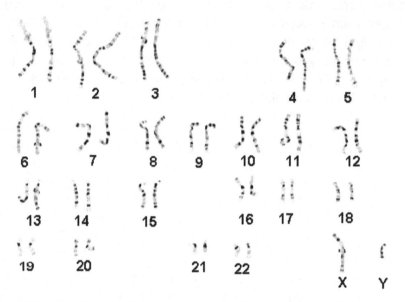

Fig. 12.1. Chromosomes from a male nucleus. Note that the male Y-chromosome (bottom right) is much smaller than the female X-chromosome (from Wikipedia http://www.chromodisorder.org/CDO/General/IntroToChromosomes.aspx).

Before mention is made of a more fundamental theory, it may be useful to try and define some run-of-the-mill biological terms now even in every-day use. We mentioned earlier (see page 32) that our body tissues are made up of cells. These are visible with a good microscope, when it is found that they contain a spherical structure, the nucleus. It can be made to stand out when the tissue is treated with a suitable dye. Finer resolution of the nuclei with an electron microscope reveals that they contain elongated twisted structures, namely chromosomes, 46 in total, i.e. 23 inherited from each biological parent (Fig. 12.1). Chemically they can be described as proteins and DNA. Each set of 23 contains one sex chromosome, termed X for the female, and Y for the male variety. When combined, XX are found in females, XY in males.

DNA is built from four, and only four, types of nucleotide, namely A, C, G, T, which stand for adenine, cytosine, guanine and thymine. (Nucleotides are chemical compounds made up of three linked structures, namely one containing nitrogen, another phosphorus, and, last, a sugar). In various combinations these so-called nucleotides form the strands of the famous double-helix, the pair of strands being similar. As is now common knowledge, the latter is what chromosomes consist of. The end of each strand is sealed by a telomere (ancient Greek for end-part).

Without telomeres, chromosomes would be liable to lose the genetic information they carry, and to fall victims to malfunction. Telomeres, then, are controllers which protect the ends of the chromosomal strands, and watch over the replication and stabilization of chromosomes themselves. Telomeres, in turn, contain chains of DNA, made up of sequences TTAGGG (see previous paragraph) in one strand, and of AATCCC in the other. As shown in Fig. 12.2, the length of the strands is liable to shorten with age. Since telomeres are only chromosomal tags, they are much shorter (~8000 base-pairs at birth) than chromosomes with some 150,000,000 base pairs. As already mentioned, it is the former number that shortens with age.

Telomere length is controlled by an enzyme (i.e. a facilitator of a chemical reaction) called telomerase: if its activity weakens, the telomeres shorten, which, in turn, may lead to the ageing of the cell of which they form the end-part. In most adult cells there is no telomerase, thus the telomeres shorten and this happens every time a cell divides. When telomeres have been reduced to a critical size as a result of repeated cell divisions, they lose the ability to divide further with the result that, for example, repair of damage to tissues slows down or becomes impossible with fatal results for the tissue in question. This is distinguished from other modes of death,

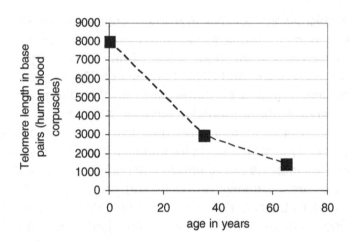

Fig. 12.2. The decline of telomere length in human blood cells with advancing years.

and called apoptosis. Cells which have lost their telomeres and stopped dividing cannot be said to have lost all function since they can switch from mediating, for example, repair to changing healthy "younger" cells into cancerous ones.

Two exceptions to this rule of decline are germ cells (see pages 50, 52), which succeed in keeping their telomerase indefinitely. The other is cancer cells which develop the pernicious ability of stimulating the formation of telomerase, thereby not only outwitting somatic cells, but also acquiring effectively immortal growth. This explains the formation of tumors.

Cells divide for growth or for repair, but the number of times they can do this is limited from 60 to 80. This is called replicative ageing, and provides an evident safeguard against cancer (see above). It suggests that human cells, or, more precisely, telomeres, can count the number of times they have divided. This mechanism is controlled by telomerase. Cells of the brain and the retina of the eye — which has evolved from the brain — do not divide, whereas, for example, those of the liver, blood and bones can do so. They are known as somatic cells, distinguished from germ (reproductive) cells which do not divide, and, in a sense, they are immortal.

The keys to ageing?

Telomeres shorten every time a cell divides, but, in *developing* young cells, telomerase manages to facilitate the restoration of their length. The shortening has been attributed to the accumulation of damage due to oxygen (see below), which may explain the difference in telomere length between healthy and damaged groups. It is interesting that ageing has been linked with telomeres on the one hand, and — in rivalry — with the effects of noxious oxygen on the other, when it is beginning to appear that the two hypotheses may be complementary. Telomeres of muscle cells form one of two exceptions: telomere length is reduced during the early years, but thereafter remains steady. This is consistent with what we learned in connection with the rate of (male) muscular ageing (see page 34). The other exception is provided by telomeres of the brain, the length of which also tends to remain constant. This, again, is echoed by a slow cerebral ageing rate in the absence of any cerebral pathology (see page 33).

The apparent link between age and telomere length seems to hold within a species, such as the human, but not between species: mice show longer telomeres than we do but have an appreciably shorter lifespan. As regards us, a study divided a selection of people over 60 years old, into two groups: one with long telomeres, the other with relatively shorter ones. The latter were three times more prone to die from heart disease, and eight times more likely to suffer from an impaired immune system than did the former: as a result, they were also much more open to fatal infections. The difference between the two average lengths of life was about five years in favour of the group with longer telomeres. It is, however, not clear what circumstances had caused the difference in telomere length between the two groups to begin with (it could be due to excess oxygen free radicals removing telomeres even when cells are not dividing). It would seem nonetheless that at least part of the genetically determined human ageing process is controlled by the telomere–telomerase complex.

Chapter Thirteen

Some Age-Related Diseases: Risk-Factors

Although ageing rates may differ from person to person and from one population group to another, ageing is nonetheless universal amongst mankind. There are, however, additional factors in operation, of which lifestyle is quoted often, and age-related diseases perhaps less so; not everyone succumbs to every age-related disease. It follows that the factors are not universal even though they may modify the ageing rate and life-expectancy. They also affect population statistics.

There is another reservation to consider. Some so-called congenital conditions may lead to symptoms later in life, and curtail it. They are excluded from our consideration because of their comparative rarity; like many others, they may occur 1 in 10,000 or less.

This is not a clinical text, but age-related diseases need to be mentioned precisely because many of them may affect individual life durations. This may be true only indirectly in connection with conditions relating, for example, to the special senses such as eyesight or hearing (which may suffer damage due to loud music): affected individuals run increased risks of accidents which may be fatal. Matters relating to lifestyle in general will be considered later.

As many daily papers keep letting us know, the effect of age-related diseases on life duration is in a constant state of flux as a result of medical advances and an increasing understanding of genetic factors. In privileged countries, previously devastating, life-threatening, infectious diseases have virtually vanished. Within living memory, breast cancer and cancer of the prostate were liable to shorten life appreciably: now it is held that early detection of either condition need not appreciably affect the duration of a patient's life. Again, "affairs of the heart" are benefiting from new medical and surgical procedures enabling patients not only to survive what formerly may have been a fatal episode, but also to enjoy more or less normal lives.

It would be pointless to try to deal with this problem exhaustively, but an (alphabetical) selection of a few well-known and frequent conditions is of interest especially because, being frequent, they can be accompanied by lists of risk-factors. These are usually categorized according to whether they are avoidable or not, a feature which makes the discussion more interesting, and which will be followed here.

Alzheimer's disease

This belongs to the dementias and is prevalent to the tune of 20–25 percent amongst those over 80 years old. Demographically, it is a matter of grave concern in view of the increasing life expectancy especially among economically privileged populations. The reason is not only that an increasing percentage of the adult population will be at risk of contracting the disease (unless some remedies are found), but also because the growing prevalence is going to place more and more stress on social services. Hard as they work, they are perennially overstretched in present-day conditions, leading to searches of solutions to problems likely to arise from a greater prevalence of the disease.

Alzheimer's disease can appear from as early as the late thirties or early forties onwards. However, patients diagnosed at an age lower than approximately 60 years probably belong to the small fraction of those who suffer from the heritable variety of the condition. By far the larger proportion is diagnosed later.

There are two genes responsible for the condition. The relatively more frequent one is a "risk gene", the presence of which increases the probability of the bearer suffering from the disease. The rarer one, called a "deterministic gene", is found in only a small number of families in the world. It is inevitably linked with the disease, and symptoms show up typically early in life.

The condition is interesting in view of the fact that a definitive diagnosis has, so far, been possible only by means of an autopsy (the surgical invasion of brain tissue, Fig. 13.1). This shows that the disease manifests in those parts of the brain that are associated with memory and with other cognitive faculties: deposits of plaques are formed and may interact with tangles of nerve fibrils.

The symptoms vary a great deal, but memory loss, accompanied by confusion, is a frequent tell-tale indicator. Often also there is present a loss of a normal, socially acceptable verbal intercourse. Furthermore, not only cognition but also re-cognition are typical features. Aimless wandering occurs frequently, and puts the individual at risk from traffic accidents. Graver variants involve a

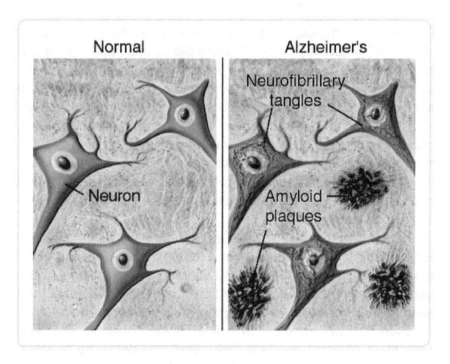

Fig. 13.1. An Alzheimer's brain cell illustrating knotted nerve fibrils (*The Alzheimer's Brain*: Christine Kennard, About.com, November 14, 2005).

tendency to paranoia and also violence. Further aggravation of the condition occurs when it is accompanied by a stroke or cardiovascular conditions.

Going, going, going, gone...

The lasso is too short to catch
The fleeing words,
Too loose to seize the thoughts
That follow them. The noughts —
Amnesia is the term — now stretch
The widening gaps in memory's net,
The net once tightly knotted of words
And knots that knitted them
Into thoughts. But not any more.

The lasso flies — in vain —
Too short however long the rope.
Again the net is stretched:
The squares that make it up
Grow into vacant voids.
Soon, soon the net will be all hole,
Without a string in sight:
The onset of an ever-lasting night.

As regards the duration of the patient's life, this seems to depend both on the age at the time of diagnosis and on gender. The balance of opinion inclines to the view that women are at greater risk than men. However, once diagnosed, women tend to live longer with the disease than men. Thus, diagnosed at the age of 70 years, women would survive half as long as equally old controls (i.e. persons without the condition), with shorter survival times for men. At the age of 85 years, the survival times are, not surprisingly, shorter, with a gender difference that persists, even though it is reduced. Further life curtailment is to be expected if Alzheimer's disease is complicated by conditions such as those mentioned above.

As far as risk-factors are concerned, they are divisible into those under our control and the rest respectively. Alternatively they can be divided into genetic and non-genetic groups respectively. It is uncertain whether the above two groupings are equivalent. For example, non-controllable risks seem to include the build-up of advanced glycosylation end products involving any type of sugar, and ultimately liable to lead to the dysfunction of affected proteins. People with a parent or a sibling who suffers from Alzheimer's disease are some three times more likely to be affected than those who have no such blood relation. Two genetic factors have been identified. One is the above mentioned "risk-gene" which leads to a possibility of the carrier being affected without necessarily succumbing to the condition. The other is more serious though, as we have seen, fortunately much rarer: it is called the familial "deterministic gene" because the result is certain: as mentioned above, the disease strikes at a relatively young age.

One of the most serious of the controllable risks is an injury to the head. This may show up many years after an accident. Boxers are in the forefront of the hazard, and the British Medical Association has been waging a campaign for many years to have both amateur and professional boxing banned in the United Kingdom. Being "punch drunk" may be the immediate effect of a blow on the head, with the irreversible damage showing up years later. The cynic may ask why rugby players should apparently be immune, until one remembers that many doctors tackle the sport in their youth.

Recently it has been reported that heavy drinkers show symptoms some five years earlier than the average of non-drinkers, with smokers doubling the risk which threatens the non-smokers. As this touches on diet, it is worth mentioning that healthy people, consuming the well-known "Mediterranean diet", experience a delay in the onset of Alzheimer's disease, and those already suffering from it may delay death by some four years on average.

Some of the less well-known risk-factors include oestrogen replacement therapy following women's menopause. While statins, and also such painkillers as Ibuprofen and Aspirin, are widely recommended because they alleviate symptoms of Alzheimer's disease, some isolated voices have mentioned them as risks, probably if recommended dosages are greatly exceeded.

There is some evidence to suggest that cognitive and mental factors may play a role in the development of the condition. For example, people who have been fortunate enough to have been exposed to significant educational influences appear to be less at risk than those who have not. Again, those who manage to be, and stay mentally agile in later years are also at a smaller risk. Indeed, a significant delay in one's date of retirement has recently been said to delay the potential onset of the condition. These factors are especially interesting because they are linked to an individual's social environment. Whether we receive any significant education or not depends primarily on our parents or guardians rather than on ourselves. And whether, to keep our mind as nimble as possible, we indulge in solving complicated crossword puzzles in our later years partly depends on the company we have been keeping.

If the story about Alzheimer's disease sounds grim, that's because it is. The suffering it entails to the individual (particularly during the early period of onset) and to his/her surroundings is sad in the extreme. But these are also the reasons why research into the disease is pushed, and pushed more, and progressing, if not in leaps and bounds, at least in gradual steps. This is only to be expected when what is involved is "the most complicated structure on Earth", namely the brain.

Common problems of the bones

Bones, assembled as the skeleton, determine the size and shape of the body, even though the final "fleshingout" is due to soft tissues. We have already mentioned (see page 24) that they have evolved along sound principles of mechanics by virtue of being strong without being solid. Bone material is being renewed throughout life. The young rate of renewal is about 10 percent every year. After the age of about 30 years, the loss of bone is not matched by the replacement process. As a result, the mass of the bone material declines, it becomes more porous, hence giving rise to osteoporosis.

It is the best-known bone condition linked to age. It has no obvious symptom, and is more a matter of "by their fruits ye shall know them". In itself it is not life threatening, but may lead to conditions that are liable to shorten the lifespan (see below). When an elderly bone fractures this would almost certainly be due to osteoporosis. Some 10 percent of people in their sixties, and 80 percent of those who are 85 years old are liable to have porous bones (Fig. 13.2). It affects mainly women, particularly post-menopausal ones, and hence points to a hormonal involvement. The special risk-factor for men is a low testosterone level, which usually accompanies the medical treatment of cancer of the prostate. However, the latter is not a prerequisite for osteoporosis developing in men. The changes in the bones appear with increasing age, partly perhaps because one's mobility tends to be reduced. It is noteworthy that, when bones are under mechanical stress, their thinning is slowed down. Hence lifestyle, i.e. using legs rather than mechanical transport, is beneficial in this as in other undesirable conditions.

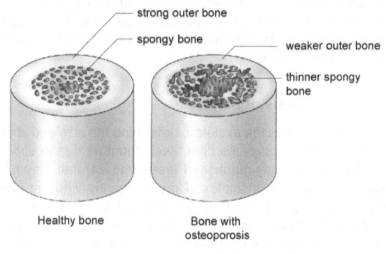

Fig. 13.2.   Courtesy BUPA.

While reduced testosterone levels affect the bones of both genders, a bye product is the reduced intestinal absorption of calcium (important for bones and transported to them by vitamin D, the "sun" vitamin). Altogether the risk of osteoporosis in men is not negligible, and amounts to 25 percent of them after the age of 40 years.

Quite apart from changes in the mineral composition of the bones (mainly owing to the loss of calcium), an increase in porosity and the resulting thinning of the inner (intracostal) network is the main risk-factor. The condition is usually successfully treatable.

Talking above of mechanical principles reminds one of the special problems facing the spinal or vertebral column. They became acute once human walking on all fours gave way to the present erect posture. Before the change, the spinal column served merely as a connector of the front and rear of the body (apart from containing and protecting extensive nervous control systems of voluntary and involuntary movements). However, after the drama of the evolution of erect posture, it also became the one and only supporting scaffolding for the upper part of the body, stretching, as it does, from the skull to the pelvis. Just think of the pressures exerted on the 24 adult vertebrae by a heavy head, a ribcage, and two arms, not to mention the soft tissues. This situation offers a rich scenario for complications in an ageing body. These may actually manifest at a much younger age: many people between the ages of 25 and 60 years experience at least transient or acute, as distinct from chronic, backpain. This emphasizes the precarious nature of the conversion of a quadruped into a biped.

If the spinal column were one solid bone, existing problems would be minimized, even though the mobility of the body would be gravely compromised. The flexibility and mobility of the column are maintained by shock absorbers between the vertebrae, known as intervertebral discs. Cumulatively, they represent about one quarter of the total length of the spinal column. When one of them has been squashed and slightly displaced as a result of prolonged wear and tear, or of a sudden disturbing movement of the spinal column as a whole, the resultant stress imposed on nerve trunks running through it is liable to cause great pain. Wear and tear throughout the years tends to compress mainly the lower discs with the result that people, who were six foot (205 cm) tall when at the peak of their adulthood, will have shrunk to five foot ten (199 cm) in later years.

Since one of the functions of the vertebrae is keeping in position muscles that are attached to them, a change in muscular tension is liable to affect the tension in the vertebral column. As, with advancing years, muscles lose the fibrils they consist of, and, with the loss, their erstwhile functional ability, the deficit affects posture. This is a contributory factor in spinal bending (dowager's hump) seen almost universally in the elderly.

As is true of other diseases, the risk-factors of osteoporosis fall into two groups, namely unavoidable and avoidable hazards. Age and being a woman belong to the former group, as does belonging to a family with osteoporosis (which suggests a genetic trait). Added to these are thyroid disorders, some types of kidney disease, and rheumatoid arthritis, the latter exercising a potentially adverse influence on the length of the lifespan. This can also be affected as an indirect result of osteoporosis when this has led to a severe fracture, and, say, pneumonia, developed in the course of what should have been a period of recovery.

The hazards under some control include smoking and a hefty intake of alcohol. Bed rest or physical inactivity also contributes to the hazard: this is especially noteworthy because walking and exercising, for example wrist activity, is known to be beneficial. In fact, most exercises involving supporting one's own weight are recommended for increasing one's bone mass. Low levels of calcium and Vitamin D are clearly remediable, the one by diet, the other by a reasonable exposure to the sun.

Another, fairly well-known, condition of the bones is arthritis. This is a condition of the cartilage, a disc-shaped connective tissue forming the joint between the ends of two adjacent bones: this facilitates movement for example of the elbow or the knee. There are some 200 types of arthritis, but the most prevalent ones are osteoarthritis, and the somewhat rarer, but much more painful, rheumatoid form.

The onset of the former condition lies in the two decades between 40 and 60 years of age: some 10 percent of women in their seventies suffer from the osteo type; it is less prevalent amongst men. As age advances, the cartilage between the bones becomes gradually ground away, causing acute pain when the previously separated bones start rubbing each other. The areas to suffer most frequently are joints in the hands, the spinal column, the knees, and the hips, in other words, the joints having to do most of the work during our lives.

The slowly developing symptoms include pain and, possibly, stiffness in the arthritic joints. Later on the joints may get deformed to a visible extent, and the pain tends to become aggravated in the course of a day as a result of movements of the joints. When the arthritis occurs in the vertebral column, it weakens and ultimately tends to bend with the shoulders dropping forward.

Although the causes of osteoarthritis remain to be established, some risk-factors have surfaced. Prime among them is obesity since, for example, the knee joints are subjected to a greater pressure than they have evolved for to bear. The suggestion that there may be a genetic element in arthritis has not been proved (though also claimed for obesity). Another cause is injury particularly of the limbs for example in the course of a game of rugby. A relatively modern hypothesis points to a repetitive exertion, such as typing, which may give rise to repetitive strain injury (RSI).

Preventive measures include the control of one's body mass, and exercising the joints at risk. It has been suggested that this should be continued even when symptoms have already staged their appearance.

The rarer condition of rheumatoid arthritis, less prevalent than the osteo type, is liable to surface between the ages of 30 and 50 years, and is attributed to a failure of the immune system. In this instance, it attacks tissues of the body which it is normally expected to protect. The cause of the condition is thought to be genetic. Women, again, are much more likely to be affected than is true of men; however, with the younger onset, no direct hormonal link appears to be suspected. The progression of symptoms is similar to that in osteoarthritis, but the diurnal change is inverted: pain tends to diminish throughout the day. Unlike osteoarthritis, the rheumatoid type may be sensitive to weather conditions, apparently being aggravated when it is wet and cold.

At present there is no known means of prevention. Early diagnosis is important since considerable loss of cartilage is believed to occur early on in the course of the disease. The alleviation of symptoms is helped by swimming because this exercises the joints without any significant pressure being applied to them. No dietary influence has been suggested so far, though fish oil has anti-inflammatory attributes. As regards, "over-the-counter" pain killers, one patient differs from another: hazardous side effects need always to be guarded against.

A rare affliction of bones, namely cancer, is mentioned in the next section 'Risk-factors for some cancers'.

Risk-factors for some cancers

There exists a large variety of cancer: this brief survey lists a few of the relatively common ones that are age-related (see Fig. 13.3). It is, of course, arguable to what extent diseases may accelerate the processes of ageing. While any categorizing is partly arbitrary, the fact remains that mortality due to cancer affects the data on life-expectancy, and they certainly influence our thinking about the (current) length of life and hence ageing. Furthermore, some of the fairly certain risk-factors can be traced to early life (a tune that will be found to be repeated in these pages), which should make the subject of interest to younger heads.

In general, the number of cases has been on the rise in most privileged countries: this is largely due to the nearly universally marked rise in life-expectancy during the last 100 years or so. There exist, however, exceptions: for example, among populations who have reduced the extent of smoking, the incidence (i.e. new cases)

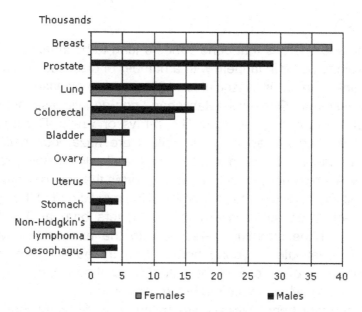

Thousands

Fig. 13.3. Horizontal axis: United Kingdom cases per thousand (http://www. statistics.gov.uk/cci/nugget.asp?id=915).

of lung cancer has slowed down. Again, the relatively large Japanese incidence of cancer of the stomach in that country has been linked to the country's diet, but it started falling once refrigerators became widely used in the second half of the twentieth century.

Another point to bear in mind is that cancers are divisible into two groups, namely primary and secondary. For example, cancer of the prostate is seen, at present, as a primary cancer. In advanced cases, the disease is liable to spread to bones as a secondary feature due to prostate cancer.

In a minibook such as this, there is no point in discussing clinical aspects. What is likely to be of interest is a listing of risk-factors such as they are known at the present time; particular emphasis will be given to those which can be individually controlled, surely a matter of special interest to a young inquirer. For simplicity's sake, the selected conditions are listed in alphabetical order.

Bone cancer

The paradox here is that bone cancers tend to be inherited. They are liable to occur in tandem with a number of diseases; for example bone cancer is associated with an eye disease, namely retinoblastoma. One well-established, non-genetic risk-factor is exposure to ionizing radiation, like X-rays or γ-rays. Radioactive materials such as radium or strontium are hazardous because they are liable to build up in bones with a slowly decaying radio-activity. A rapid decay would evidently shorten the period of hazard. The test explosions of atomic bombs in the middle part of the twentieth century caused concern for youngsters born at that time because of the strontium released into the atmosphere which would accumulate in young bones. Primary bone cancer data exclude those of cancers of the soft innards of bones, namely the bone marrow, which forms red blood corpuscles.

Secondary bone cancers are much more frequent than are primary ones, but, as we saw above, appertain to their primary progenitors of which cancer of the prostate is a prime example.

Brain cancer

Here again, there are two types, namely primary tumors, and secondary ones, due to cells that have migrated from other affected parts of the body. Brain tumors are frequently fatal, as they interfere with the crammed space taken up by important control organs in the brain. This is as true of non-cancerous tumors as of cancerous ones.

There are a number of brain tumors resulting from genetic causes. Rare cases of tumors of the brain and the spinal chord (an extension of the brain) can be inherited. With the exception of radiation, and a chemical, vinyl chloride, there is no known chemical or environmental hazard leading to tumors of the brain. Exposure to radiation may not manifest adverse results until 10 to 15 years later. Vinyl chloride is used in the manufacture of plastic products employed in the home and in some car components. The users of the plastic items are not at risk, though this is less true of those involved in their manufacture. Vinyl chloride is present in tobacco smoke.

Disorders of the immune system represent another risk-factor. This may be due to congenital causes, or to lymphomas, originally forming in lymph nodes (for example in the crotch). A defective immune system with risks for the brain may also result from AIDS.

Breast cancer and ovarian cancer

Breast cancer is the most frequently encountered cancer, affecting women about 100 times more than it does men. It is relatively rare amongst women aged 45 years or less, but increases in frequency amongst those aged 55 years and over, i.e. as a post-menopausal feature. Between one in 20 to one in 10 cases are thought to be inherited, and to result from mutations of one of two genes BRCA1 and BRCA2 (the acronyms stand for breast cancer). A handful of other genes promoting cellular growth and suppressing tumors can also be involved but to a smaller extent than the previous two.

The hazard to women resides in their breast cells being constantly exposed to two growth controlling hormones, namely oestrogen and progesterone. A family history of breast cancer raises the risk, even if male rather than female relatives are involved. One affected breast raises the odds of the disease occurring in the other.

Although white women are at a somewhat greater risk than are Afro-American, the condition is more likely to be fatal amongst the latter. However, Asian, Hispanic and native American women are at a smaller risk both of developing the disease and of dying from it.

The age of onset of menstruation (before 12 years) and that of the menopause (after 55 years) increase the risk. However, as these thresholds tend themselves to vary with environmental influences, such as diet, ambient temperature and social circumstances, the above age points may not be basic determinants.

As might be expected, irradiation for other purposes, e.g. lung examination, significantly increases the risk; as in many other instances, a judgment has to be exercised as regards the relative risk magnitude. This is also true of medication, such as hormone replacement therapy (HRT), which may have the risk of breast cancer associated with it as a "side effect". Incidentally, an important measure under individual control is screening, both by self-examination and clinically. This also applies to cervical cancer, not further addressed here.

Hazards more under the individual's control include the age of having children and their number: more than one pregnancy at a younger age tends to reduce the risk (but see below). Post-menopausal hormone therapy (PHT) increases the risk of the disease, and of dying from it, but the odds return to those of the population at large if the PHT has been avoided for a period of five years. Young mothers, breast feeding their offspring, particularly for a lengthy period, tend to benefit from a reduced risk. This is also true if they practice regular physical exercise.

Alcohol, which promotes cancers of the mouth, oesophagus, and liver, raises the risk of breast cancer by 50 percent for those who have two to five drinks every day. Overweight and obesity are interesting in that they raise the odds primarily when they occur in middle age: obese teenagers would appear to be at a relative advantage over more mature individuals.

(The above factors have been established with a fair degree of certitude. Less reliably investigated risk-factors have been published, but they would be considered only if confirmed.)

It seems reasonable to consider ovarian cancer in the present section since the two share some important risk-factors. The risk decreases with each pregnancy. But overall, the incidence of ovarian cancer increases with age, as does the resulting mortality. There is a statistical 10 to 15 years' delay between diagnosis and demise. Obesity is a risk-factor, increasing by 50 percent amongst the heaviest women. The hazard is also increased by oestrogen replacement therapy, as it is by a family history of breast, ovarian and colorectal cancers. Earlier breast cancer also constitutes a risk. The risk is also increased if a close relative, aged 50 years or less, has had ovarian cancer. There seem to

"If you give up alcohol, cigarettes, sex, red meat, cakes and chocolate, and don't get too excited, you can enjoy life for a few more years yet."

Fig. 13.4. www.CartoonStock.com.

exist also dietary influences: a prolonged low-fat diet is benefi-cent, as are "healthy" vegetables and fruit. As appears to be true of cancer of the prostate, a high intake of red meat (Fig. 13.4) and of processed nutriments is to be avoided.

Colorectal cancer

Figure 13.3 shows that colorectal cancer is relatively very rife, although with improved hygienic conditions in relation to the sale and storage of food products it has shown a decline during the last few decades.

The comparatively large incidence of colorectal cancer is partly due to a substantial number of risk-factors, both inherited and those that can be controlled. Chief amongst the former is polyps of the colon. A treated episode of colonic cancer, even if treated successfully, also constitutes a hazard for repetition. As so often, family history also raises the odds. However, this is true of only 20 percent of cases; most cases are "new". The 20 percent include a number of familial and hereditary syndromes, some of which manifest when the individuals are relatively young. Some of them put women at risk of developing cancer of the endometrium, i.e. the lining of the uterus. A previous case of inflammatory bowel disease constitutes a risk, but has to be distinguished from irritable bowel syndrome which does not do so. In general, any trace of blood associated with waste products needs investigating.

Ethnicity appears to be playing a role. African Americans are highly susceptible for reasons unknown, whereas genetic causes have been established for Jewish individuals of Ashkenazi (East European) descent; they are also highly at risk.

Amongst the controllable causes of colorectal cancer, smoking ranks high: it is thought that cancer-promoting substances get swallowed, and are therefore enabled to act directly on the colon. The dietary strictures mentioned in the previous section are held to apply also here. In addition, modes of preparation of meat involving high temperatures, as for example, in frequently repeated barbeques, may lead to the production of carcinogenic chemicals, again raising the odds. Views on alcohol consumption differ, but the balance of opinions is summarized by "better safe than sorry". Lack of physical activity, accentuated by obesity, is amongst the most potent risk-factors for this cancer amongst most others.

Since diabetes type 2, i.e. non-insulin dependent, may be partly self-inflicted, mention should be made of the fact that it is a risk-factor for colorectal cancer. This is so, even if diabetes or the other risk-factors have been allowed for in the calculation of the hazard.

Lung cancer

This type of disease is unusual in that most of the risk-factors are of an avoidable nature. Lung cancer has received a great deal of publicity in the popular press, largely because it is linked to a care-fully quantified risk-factor: the more one smokes the greater the probability of the development of lung cancer. Smoking is a pene-trating risk-factor, linked not just to cigarettes, but, to a smaller extent, also to cigars and pipes. It is important to note that non-smokers exposed for prolonged periods to a smoking atmosphere are also at risk. Fortunately, an abstention of 15 continuous years reduces the risk to its normal (non-smoker's) level. In addition to tobacco, cannabis smoking constitutes a risk. The occurrence of the cancer in non-smokers is rare. It should be mentioned paren-thetically that the effect of an abstinence from alcohol is harder to evaluate since the consumption is unlikely to be quantified with the precision possible for smoking.

Two environmental factors have been reliably identified. One is exposure to asbestos. However, this has been under strict control in the United Kingdom for many years so that the actual cases are relatively rare. The other is exposure to radon gas which may, in some locations, have a radioactivity significantly above that of normal surroundings.

Cancer of the prostate

This condition is among the most widely spread among cancers. Rare below a man's age of sixty years, it reaches about 80 percent of Western men in their early eighties. There exist variations due to ethnicity: thus, in contrast to the rare testicular cancer, Afro-Caribbean men have a higher prevalence than White men, whereas Asians exhibit a lower one. However, when Japanese, who have a low prevalence in their homeland, migrate to the United States of America, the prevalence of their condition proceeds to approximate to that of the host inhabitants. This points to the conclusion — one that is reassuring in the long run — that environmental conditions rather than genes appear to be a dominant risk-factor. For example, it has been suggested that a diet low in fruit and vegetables, but very high in animal fat — which would include dairy products — increases the hazard. Dairy nutrients constitute a risk also on account of their calcium content. The hazard is, however, said to be reduced by the consumption of selenium and Vitamin E and, more especially, lycopene which is contained in tomatoes and tomato ketchup, and is a powerful antioxidant. However, reports on the matter are mixed in their recommendations, and no critical clinical assessment appears to have been published to date. Moreover, like so much dietary advice related to health, no quantified data appear to have been published to buttress the pieces of advice on offer.

Family history is, again, an important risk-factor, and especially interesting in that it involves women in addition to men. Thus, if one side of the family contains patients with breast cancer, particularly if developed at an early age, the risk is increased. If a close relative like father or a brother has developed the disease under the age of 60 years, the risk is greater than it is with a more distant male relative, including uncles and grandfathers.

Like breast cancer, that of the prostate is thought to involve hormones. The male hormone testosterone stimulates the growth of both healthy and cancerous prostate tissue: one of the therapeutic procedures, currently in use, is directed at a reduction of its concentration. Smoking, which we have seen to be a hazard in other respects, has also been mentioned as increasing the risk of prostate cancer. If not detected in its early stages, prostate cancer is associated with a considerable mortality, largely because of the spreading (metastasis: transfer) of the cancerous cells to form secondary conditions.

Though at present partly controversial, screening is known to have prolonged the lives of a significant number of patients diagnosed with cancer of the prostate.

Stomach cancer

As mentioned earlier, one of the most affected countries, Japan, has recorded appreciable improvements in the prevalence of this condition, probably owing to improvements in diet hygiene, and also some in preventive diagnostic advances. Even so, not so long ago, the risk in Japan was still about three times greater than in the United Kingdom.

The condition occurs more frequently in men than in women, and tends to peak before the age of sixty years. The causes are still being debated. Genes seem to play a role. While food that has been pickled, salted, or smoked increases the risk, Vitamin C, fresh vegetables, milk, and also certain types of frozen food appear to be beneficial (Fig. 13.5). Some uncontrollable factors, namely

"My great granddad says fruits and vegetables keep him healthy. He calls them 'fossil fuels.' "

Fig. 13.5. www.CartoonStock.com.

belonging to the blood group A, and housing bacteria of *Helicobacter pylori* in our stomach, increase the hazard. This is also true of a variety of preceding diseases of the stomach, particularly those affecting its lining. In the United Kingdom, the mortality due to stomach cancer occupies the seventh place amongst the major cancers.

The heart at risk

Though widespread in privileged countries, affairs of the heart are subject to a large number of risk-factors. Quite a few of them are open to modification. This does not mean that following the advice offered (Fig. 13.6) eliminates the hazard: it merely changes it on a statistical basis.

Remember that statistics are blind to individuals; only masses count. This makes it possible to determine the odds with which an individual runs a risk of developing a problem. It is generally thought that a risk of 20 percent or over is high, whereas one of 10 percent or less is low. What this means is this: of 100 patients sharing the "high risk" characteristics, about 20 will succumb, though there is no knowing which 20. Similarly, of a 100 in the "low-risk" category, about ten, but an unknown ten, are likely to develop the condition. There is, however, a possibility that the risk calculation may lie on the pessimistic side. The reason is that the incidence (number starting) of several conditions is improving so that the well-established markers may be getting out of date. For example, myocardial infarction, the irreversible death throes of part of the musculature of the heart resulting from a closed blood supply, is on the decline in the United Kingdom; and heart disease, more generally, is becoming much less frequent in the United States of America.

Most of the problems in the category of heart disease, and in that of stroke, arise from a plumbing problem: arteries (blood vessels) ferrying blood from the heart, and also maintaining the function of the heart itself, tend to have their calibre restricted, if not altogether blocked. This reduces the rate of supply of essential nutriments, and ultimately interferes with function.

Lower Your Blood Cholesterol

Learn Your Cholesterol Number

Cut Down on Fat— Not on Taste

Watch Your Weight

Prevent High Blood Pressure

Cut Down on Salt and Sodium

Stay Active and Feel Better

Kick the Smoking Habit

Fig. 13.6. Tips and warnings (http://www.nhlbi.nih.gov/health/public/heart/other/sp_smok.htm).

The blockage is due to atherosclerosis, or "hardening of a paste"; in this case the paste is made up of fatty deposits. These, in turn, are usually the result of an excessively fatty diet.

As with most other age-related conditions, risks are divided into musts and mays. The former are unavoidable, while the latter may be under the patient's control.

It is interesting that family history has been included in the former category without any responsible gene having been identified. When we consider that diet is largely a matter of the culture one belongs to, the technical concept of the familial influence becomes tenuous, since it is amenable to change. Other unavoidable risk-factors include being a male, or having an early menopause, or being a Bangladeshi, Indian or Pakistani. This means that an even greater effort should be devoted to minimizing such modifiable risk-factors as may be relevant.

The modifiable causes have been well publicized in the daily press. They include smoking; with the widespread use of cigarettes, they are the main target. But, as previously mentioned, hot items of food ingested into the body may become carcinogenic no matter what their origin. A sedentary life style is another serious risk-factor, because the heart is not subjected to the exercise which, for example, a smart rate of walking may provide.

If this is combined with an "unhealthy" diet, and one including an excessive consumption of salt, it may lead to an individual becoming overweight or obese, which is a risk in its own right. An over-indulgence in alcohol is also a risk-factor, as suggested by Scottish statistics on the prevalence of heart problems.

A comment on the matter of someone being overweight is apposite at this juncture. A person can be "overweight" owing to one of two causes: s/he can have a surplus of fat or, alternatively, be very muscular. The former condition is liable to be included with risk-factors. However, the latter appears to be beneficial. Indeed, a detailed study reported in 2006 that being overweight, as due to a strong musculature rather than to the accumulation of fat, is liable to increase life-expectancy.

There exists an additional group of partly controllable risk-factors in the form of disease conditions. These include high blood pressure, which tends to be controlled better in privileged societies than for example in poverty-stricken Africa. A high cholesterol level, partly due to diet, and partly determined by genetic factors, also raises the risk. A high fat intake is another risk, as is diabetes. The latter straddles unavoidable and avoidable groups of hazard, as do a number of diseases of the kidneys. Diabetes is a marked contributing factor to the mortality rate of heart conditions. Although the life-expectancy of patients with diabetes mellitus is approximately one-third less than that of the general population its influence on other conditions, such as the above, depends on its duration. For, conversely, heart and brain conditions are the common causes of death in diabetics.

The mortality effect of heart diseases does not need to be stressed: few are the people who have not heard of someone in their circle dying from a heart attack. The mortality rate has fortunately been decreasing during recent decades in certain geographical regions.

Parkinson's disease ('the shakes')

This condition is one of the commonest neurological diseases to befall the elderly. It has been estimated to affect some 2 percent of those in their early eighties, with the proportion rising, being strongly age-related. In residential and nursing homes the prevalence may be as high as 10 percent.

A degenerative and, at present, unfortunately incurable condition, its symptoms (Fig. 13.7) are partly amenable to treatment and alleviation. It is, however, an especially hazardous condition because it is associated with falls, fractures, and other problems, leading to relatively frequent hospitalization. It affects men and women alike. It is not easy to diagnose: errors of up to 50 percent have been reported. This makes it difficult to accept as reliable, reports that there exists a geographical variation of the prevalence of the disease. Cognitive impairment has been reported in connection with the condition. Unlike problems of the heart, Parkinson's disease shows hardly any association with avoidable risk-factors. The unavoidable ones include a dozen or so mutations of genes, which are inherited. Some may be due to the environmental presence of toxins. An important non-genetic risk-factor, though it may not be avoidable, is trauma to the head: severe injuries to the skull show a significant correlation with the incidence of the disease.

While the condition itself is not considered to be fatal, its consequences, such as pneumonia and dangerous falls, for example,

Fig. 13.7.   Illustration of Parkinson's disease by Sir William Richard Gowers from *A Manual of Diseases of the Nervous System* (1886).

are liable to curtail life. One study has reported a greater tendency for men to die of the disease than is true of women. Also (American) non-whites were at one-third the risk recorded for whites.

Stroke: the brain at risk

A stroke is the result of an impaired blood supply to the brain. The uninterrupted supply is crucial to survival: in extreme cases, a cut-off of four minutes' duration can lead to irreversible damage, if not a fatality. Like other malfunctions resulting from an impaired blood flow, strokes are age-related: the majority of them occur after the age of 65 years. Broadly speaking, there are three major types of stroke: (a) the ischaemic (bloodless) type, resulting from a blockage of the blood supply to the brain, (b) the haemorrhagic type, resulting from bleeding within the skull, and (c) the transient ischaemic attack (TIA or ministroke), which is relatively harmless, its symptoms usually disappearing within a day.

Although strokes are more numerous in men than women, they are more likely to prove fatal in women than in men. There exists also an ethnic extension: African Americans are more at risk than is true of Caucasians; it is an open question to what extent this may be the result of a parallel imbalance of some of the risk-factors listed below.

There are numerous symptoms associated with each of the above types, depending on the precise location of the incident in the brain. In general, the risk of a stroke increases when there has been a previous episode. Many of the risks listed in connection with heart diseases apply also here. Incidentally, a previous heart attack constitutes a risk-factor for stroke. As is to be expected, uncontrolled high blood pressure (hypertension) is another. Diabetes, a high level of cholesterol, and a family history of stroke also figure in the list. As regards the last, this does not necessarily imply a genetic involvement; since dietary aspects may also play a role, these may be a matter of habit as between one generation and the next rather than a case of inheritance of genes.

The Stroke

The baton tip quivered in the air.
Tense, definite, minute and all-important,
The centre of the universe of every pair
Of eyes. Then, for a timeless instant,
It stood still,
And every lung heaved heavier till —
The instant over — the tip descended.
The baton followed into the depth. The hand
Pursued it, raised it, slowed it, stopped it.
The strings drew a barely audible note and
The drums, on whose crescendo the effect depended,
Rolled from the distance like a plane.

Preceding the awful cataclysm by a tenth of a second,
The tails flew into the air, and, as if jabbed
By the baton, all but collapsed on the floor.
The roar
Of the elements was unleashed, and ebbed
Only to strike the ear-drum with sonorous surprise.
The man rowed the boat and rode the storm.
He pleaded with plaintive flutes
And made the bassoons confess
To base undertones; then his wand
Forced human voices from the elephantine shapes
Of the cellos. He soothed the trumpets,
Roused the dreamy clarinets.
Ah power, power, thrice blessed power.
The gates were open, but the flood was benign
And rose and fell with every sign
From the circulating tip.
It is a pencil and, as the music streams
From the loudspeaker in (mildly) distorted strains,
He follows it closely. Sometimes the lip
Helps where he dreams,
And delayed atonal gurgles
Emerge from deep down.
It might have been me. The thought burgles
The mind, and the pencil, egged on with a frown,
Belatedly beats the fleeing rhythm.
The temporal veins
Excresce, as the mighty sforzato
Is struck by a vacuum.
The music plays on. The music plays on:
Kyrie eleison.
The flutes complain no more:
The base undertone dies
Since the pencil lies
In splinters on the floor.

At the risk of being repetitive, we find that the wrong type or amount of food is a risk-factor also in connection with strokes. This includes an excess of saturated fat, a lack of fruit and fresh vegetables, and an excessive devotion to alcohol. Smoking and drugs, notably cocaine, also figure in the list of risk-factors. The potentially resulting obesity and lack of physical exercise (cause vs. effect?) also raise the odds against a life being long.

Chapter Fourteen

The End of Ageing

We saw earlier that it was Cicero who noted that ageing — according to him, a disease — creeps on us stealthily, but finishes unambiguously. Although death certificates usually mention only one cause, probably the final or immediate one, the odds are, if there had not been one cause, there would have been another (it has been estimated that fatal accidents involving cars present with six-fold overkills!). When the great playwright George Bernard Shaw died at the then ripe age of 93 years, the cause of death was popularly held to be "old age". Nonsense — he died of pneumonia; contrary to Cicero, old age is not a disease but a great facilitator for its occurrence: anyone who calls age a risk-factor may be thought to confuse cause and effect. In other words, risk-factors for terminal diseases become more numerous, or more and more effective, or both, as the years advance. At the same time, our immune system, which protects us well from a variety of conditions during our first five decades or so, itself starts to falter, and allows hostile elements free access to our body. This is one way of explaining why an elderly person who has suffered a fall and broken a hip may die of pneumonia. The pneumonia is not a consequence of the accident; but, the fact of being bed-ridden coupled with reduced mobility is liable to lead to an increased exposure and sensitivity to noxious agents.

We mentioned earlier (see page 33) that the efficiency of many biological functions drops notionally to zero in the region of a lifetime lasting some 120 years. It can be argued, of course, that effective failure might occur when the efficiency has dropped to ten percent or some such value rather than zero percent. This, however, is a detail. No one has measured grip strength (see page 26) with a 110 let alone a 120 years old person: the ages for zero values are determined from trends in younger years. It is the extended ranges that tend to congregate near an age in the 120s.

It is interesting to compare this value with mortality data for the year 2002 (for the United States of America). Mortality is a ratio defined as the number of people of a given age who have died at the end of a year, divided by the number of equally old people alive at the beginning of the same year. To make the numbers manageable, the ratio in Fig. 14.1 is given for 100,000 persons alive at the start of the year. Note also that the number of dead (along the vertical scale) is expressed logarithmically; thus the orders of magnitude rise in equally large steps.

Now the equation inserted in Fig. 14.1 shows that the age at which the death toll reaches 100,000 out of 100,000, that is to say

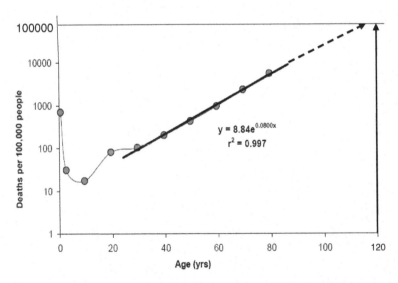

Fig. 14.1. Mortality rates, expressed in deaths per 100,000 people, as a function of age for the 2002 US population. The dotted line represents the Gompertz function extrapolated from the mortality rates after maturity. It reaches the 100 percent level for an age of ~120 years (after: CDC/NCHS, National Vital Statistics System, Mortality).

for all to be dead in the course of one year, occurs with that particular population at 116 years and eight months. When we consider that there is no information on the spread of the data as would be the case if the counting were repeated with another, but comparable population, the above numerical result is not far removed from the predicted functional failure of about 120 years. This gives us a numerical relation between the processes of ageing and its termination.

Whether desirable or not, the result suggests that an extension of the human life-span beyond this verified number of years would probably necessitate not just a means of extending the years but also a fundamental restructuring of our biological make-up. We noted earlier that an extension would hardly be welcome unless the hidden threat of unknown diseases could be tackled simultaneously. The next section may help to indicate what may perhaps be of some use in the existing state of affairs.

Chapter Fifteen

What Can We Do About All This?

Fig. 15.1. Michelangelo's *Cumaean Sybil* on the ceiling of the Sistine Chapel, Vatican (www.fisheaters.com/sybils.html).

The age-long quest for eternal life is typified by the mythical Cumaean Sybil who was granted the fulfilment of one wish: she chose eternal life, which she was doomed to spend in a haggard manner because she had not wished for eternal youth instead (see her portrait by

Fig. 15.2.   Choose your parents correctly.

Michelangelo in the Sistine Chapel in Rome, Fig. 15.1). Many of us are not very concerned about eternity, and would probably opt for long health. The standard advice is: "If you want to live a long and healthy life, you have to choose the right parents" (Fig. 15.2).

However, if that is really our wish, choosing the "right parents" is not the only option available to us. It is common knowledge that the control of our lifestyle lies at least partly in our hands. There are a number of important non-scientific points to clear up on this subject. Providing information — the best information available at the present time — does not imply that its purveyor intends to be prescriptive in any sense of the word. If information on a subject is available, one takes it or leaves it. The second point is that the mere consideration of new information on lifestyle and/or diet offers a challenge to established habits. Mention was made earlier that what we like to eat is often the result of what we were given to eat when we were small children. In other words, most nutritional tastes are acquired. This is as true of sweets as of the alcoholic drinks we were introduced to somewhat later, in fact of most habits which contain an addictive element.

There is, in addition, a psychological side of information relating to the above subjects. It is purveyed by "them" to "us". And "us" are frequently unfamiliar with the detailed reasons for the professional advice being proffered. It is, of course, incumbent on the "them" to be as lucid as may be possible. But some sort of barrier — sometimes a social one — needs overcoming, and that requires an effort. Another matter relates to the validity of the information on offer. Unlike lawyers, scientists and doctors can never offer any information that is final. As mentioned above, the information can be only the best available at any given time. Not so long ago dieticians advised against the hazards of potato consumption. Then, after new research, it turned out to be good for you. In other words, we are offered not the last word but the latest one.

The above points need to be borne in mind in the following (non-prescriptive) pages. They contain some of the relevant information, at present available: it can be taken, or left alone.

Smoking and the eyes

Although the relation between mortality and eye diseases may not be obvious, when they lead to impaired vision they may also lead to accidents, which, as we learned earlier, may involve fatal complications. It seems to follow that controlling eye problems as far as this may possibly, in a sense, delay ageing.

While the risk of developing lung cancer as a result of smoking is well known (Fig. 15.3), its influence on eyesight is universally unfamiliar. It constitutes a hazard in two parts of the eye, namely the lens and the retina (Fig. 9.1). The transparency of the lens is sometimes impaired in later years as a result of developing a cataract. The chance of this happening is increased by a factor of two or three as a result of smoking. Suggestions for the mechanism involved have been offered. For example, analysis has shown that cataractous lenses

© Mike Baldwin / Cornered

"But, can you help me quit smoking?"

Fig. 15.3. (www.CartoonStock.com).

may contain copper, lead, and cadmium, which may have reached the lens as a result of the inhalation of tobacco fumes. While it is true to say that cataract has lost its terror, at least in well-off countries with great surgical facilities, when the lens is removed surgically and replaced by an implant, a surgical invasion of the body always carries a risk however small it may be. Recourse to pipe smoking in order to avoid cigarettes is not advised since the risk of a central (very noxious) cataract is even greater than results from smoking cigarettes.

Alcohol consumption tends to reduce one type of cataract (cortical), and this is linked with heavily drinking current smokers, consuming at least four drinks a day, but not with those who have never smoked. It is obvious that this remedy will be outweighed by other undesirable results.

The surgical salvation is not available to those whose smoking habit accelerates the development of age-related maculopathy (see page 40). The prevalence of this condition rises appreciably after the age of 65 years, and smoking increases its hazard two- to threefold. Cessation of smoking for ten to 15 years reduces the hazard to that for non-smokers.

The mortality associated with the above two conditions has been studied. Cataract seems to hive several years off the lives of women if they have given birth to three or more children; the mortality of men appears to be unaffected. A similar gender difference exists in the relation between maculopathy and mortality. However, in the above, and also other conditions it is important for us to be aware of links that may have developed earlier in life. For example, obesity is a risk-factor for age-related maculopathy in men, and obesity is liable to reduce the life-expectancy of both men and women. Thus the ocular condition is more likely to be aggravated when accompanied by obesity than when weight is normal.

Returning to smoking, the virtually unanimous consensus of eye specialists is that smoking is an unquestionable risk-factor for the above, and also some other, eye diseases. There is also agreement on the observation that the results are not inevitable, because it has been found that those who have given up smoking for the above-mentioned weaning period of ten to 15 years will ultimately be exposed to no more than the normal risk.

Can mental attitude play a role?

This may be a peculiar question to ask in a text that deals predominantly with the body rather than the soul. However, another factor which has been shown to link up with a prolongation of life is religious activity. Although a group of nuns have been shown to live beyond their ninetieth birthday without developing Alzheimer's disease, full-time religious activity has limited appeal. However, the mere weekly attendance at a religious service appears to correlate with an extension of life of between two and seven years. In fact, the subject has been reviewed in a major analysis, attention having been concentrated so as to deal with so-called confounding factors. These are positive or negative influences, such as education, wealth, intake of medication etc. which might distort the effect of religious activity because religious people are said, for example, to tend to smoke less and to consume less alcohol than non-religious ones.

An analysis of 41 studies, involving well over 100,000 participants, came to the conclusion that the bias in favour of religious influence on the length of life is more pronounced in those tests where women are dominant. This might be explained on the basis that women tend to be more religious than men and that they have the longer life-expectancy. As correlation offers no proof of causation, one could argue that women live longer because they are more religious than men. Accordingly, the authors of this analysis stress the need for research workers to exert strict controls of confounding factors precisely so as to ensure that potential paradoxes, such as the one suggested in the previous sentence, are avoided.

What else can be done?

The word "lifestyle" has turned up in this text nine times; "(physical) exercise" nine times; and "diet" 46 times. Some repetition, you may say. The fact, however, is that our diet forms part of our lifestyle, and that the comparison of the diets of various populations in the world has been held to provide useful pointers. For example, the inhabitants of the Pacific island of Okinawa have the longest life-expectancy in the world, and it is comparatively free of disability (Fig. 15.4). They have the largest number of centenarians in the world (their proportional number is four times larger than anywhere else). They are largely healthy; for example heart disease, breast cancer and cancer of the prostate are extremely rare amongst the older generation. However, recent research has shown those younger Okinawans who are eating foods preferred, for example, by Americans, are developing health problems associated by some experts with American diets. The "life-extending" Okinawan diet offers an admirable example of diet restriction, more as regards mass than variety. The standard includes a daily seven helpings of vegetables and a like number of grains, i.e. bread, noodles and rice. In addition there are three times a week two to four servings

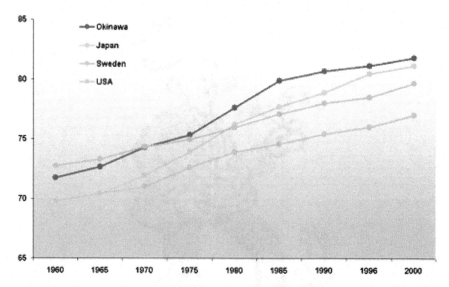

Fig. 15.4. Life Expectancy in Long-lived Populations and the US.
Source: W.H.O. 1996; Japan Ministry of Health and Welfare 2004; US Department of Health and Human Services/COC 2006.

of fruit, tofu (bean curd) and green tea. All these add up to seventy-two percent of their diet by weight, with another fourteen percent taken up by soy (beans, or milk made from soy beans) and seaweed.

In contrast, eggs, meat and poultry take up a mere three percent, while fish, rich in omega-3 as much as fourteen percent, that is to say one seventh of the average diet by weight. The overall emphasis is on balance (Fig. 15.5) including green vegetables (which contain antioxidants, remember? see page 52). Just like other Japanese, Okinawans go slow on dairy products. Note the overall absence of fat consumption. Their alcohol consumption is not zero, with men downing on average two "units" per day, and women half that amount.

A diet such as the one above may not seem very exciting as compared with average western menus. However, as mentioned earlier (see page 90), what we consume is basically the result of what we have been brought up to eat when we were very young.

"... emphasis is on balance."

Fig. 15.5.   (www.CartoonStock.com).

Admittedly, tastes can change very gradually. When we recall how many of the conditions mentioned earlier share a considerable number of risk-factors — smoking, diet, inadequate physical exercise — then the question of what can be done about it all is relatively easy to answer.

What is needed if the odds on prolonged health and also potentially avoidable disability are to be maximized is an effort of the will. This is going to be dictated by a free choice between what-I-like and what-the-current-evidence-suggests-is-good-for-me.

Unfortunately we have not evolved like rats. During the last century, a group of these scavengers were allowed in an experiment to choose nutrition from seven dishes, each of which contained a different meal. The animals were found to select, and like best, that diet that was actually biologically best for them. No wonder that rats are tipped to outlive us in the event of some dreadful cataclysmic catastrophe overtaking the world...

Chapter Sixteen

Summary of Chapters 1–15

Ageing is, at present, unavoidable. The truly unalterable component is controlled by genes.

What we can control in part is the rate at which ageing occurs in ourselves. If we decide to make the choice mentioned in the previous section, namely what-is-agreeable as against what-they-say-is-good-for-me, the choice is better done sooner than later, before one could remotely be called old (Fig. 16.1). Virtually all the risk-factors which we have discussed are cumulative, with the results of their pernicious influence taking years to become apparent.

Although "better late than never" has some beneficial effects, "early better than late" seems to have marked statistical advantages. The gamble depends, of course, on whether you believe that the statistics apply to YOU. When Cassius exclaims, in Shakespeare's *Julius Caesar*,

> "The fault, dear Brutus, is not in the stars
> But in ourselves, that we are underlings"

He seems to have been onto something...

Ageing (leading to some detailed reading)

"A certain Freed-man of Cicero's is reported to have said at a medicinal Well, discovered in his Time, wonderful for the Virtue of its Waters in restoring Sight to the Aged, That it was a Gift of the bountiful Gods to Men, to the end that all might now have the

Fig. 16.1. (www.CartoonStock.com).

Pleasure of reading his Master's Works. As that Well, if still in being, is at too great Distance for our Use, I have, Gentle Reader, as thou seest, printed this Piece of Cicero's in a large and fair Character, that those who begin to think on the Subject of OLD-AGE, (which seldom happens till their Sight is somewhat impaired by its Approaches) may not, in Reading, by the Pain of small Letters give the Eyes, feel the Pleasure of the Mind in the least allayed."

Marcus T. Cicero's Canto Major, presented by Benjamin Franklin

Chapter Seventeen

Old Age

You may rightly consider it a little peculiar that a discussion of how to delay the ageing process should be followed, rather than preceded, by one of ageing itself. However, the logic in the sequence is simple. Delay has a greater immediate appeal than what may appear to be just theory and academic information. Well then, in order to justify the title of this text, namely a consideration of factors conducive to the prolongation of youth, there ought to be a follow-up on old age irrespectively of whether it is delayed or not. Old age has been adumbrated in the Old Testament of the *Bible* (see below). Moreover, the notion of old age and its problems, diagnosed or just listed, has been written about at least since Roman times, notably by Seneca and Cicero: the latter called old age a disease: *Senectus ipse morbus est.* More significantly and lacing his view with some dry humour, he said that its onset is variable but the end well defined.

As might be expected, during most of the 2000 years since then, old age was described on and off though its symptoms started to be dealt with only once an understanding of the underlying causes was beginning to be acquired. Even so, the great Renaissance painter and polymath, Leonardo da Vinci (1452–1519) had a view to offer which could almost be part of "youth prolonged":

"Learning acquired in youth arrests the evil of old age; and if you understand that old age has wisdom for its food, you will so conduct yourself in youth that your old age will not lack for nourishment".

Again, Charles Darwin's view on ageing, such as they were, appears to have been tinged with the brush of sexism. Thus the creationist Jerry Bergman (1994) wrote:

"The racism of evolution theory has been documented well and widely publicized. It is known less widely that many evolutionists, including Charles Darwin, also taught that women are biologically inferior to men. Darwin's ideas, including his view of women, have had a major impact on society. In a telling indication of his attitude about women (just before he married his cousin, Emma Wedgwood), Darwin listed the advantages of marrying, which included: ". . . constant companion, (friend in old age) who will feel interested in one, object to be beloved and played with — better than a dog anyhow — Home, and someone to take care of house . . ." (Darwin, 1958)".

We shall see later, that, as far it goes, the current view is that women are the biologically fitter sex.

The study of ageing, gerontology, appears to have taken off seriously after the Second World War. On the one hand, there was a desire to determine how the health of Japanese survivors had been affected by their proximity to the explosions of the two atomic bombs (Hollingsworth, Hashizume & Jablon, 1965). On the other, there appeared ground-breaking texts, such as *The Biology of Senescence* by A. Comfort (1962) which showed that ageing is to be found amongst the animal kingdom in general, and not only just among humanity. Even in those more recent years, ageing would be described rather than explained. Thus an organism was deemed to be ageing if (a) it lost its reproductive faculty; (b) its sensitivity to environmental changes declined; (c) its ability to move declined; and (d) its nutrition became impaired (see Strehler, 1962).

Like most biological sciences, gerontology took off after the genetic discoveries of the nineteen fifties, when biology became molecular (see Russell, 1988). In doing so, we learned to distinguish ageing at the cellular level from ageing at the organ level, and this, in turn, from ageing of the organism as a whole.

We met ageing at the cellular level earlier on, for example when discussing telomeres, the attachments to chromosomal strands (see page 55). Ageing at the organ level we encountered several times: for example in connection with grip strength (see page 25) and also with focusing of the eyes (see page 36). And the ageing of the organism is the result of its component changes.

Cellular and organ changes have led to an interesting concept, namely biomarkers. This is the name given to functions which are characterized by a significant and systematic change with age. They can be functions that rise or fall; it is immaterial which, as long as they change progressively. An example of a riser is the hardening of the walls of blood vessels. The loss of bone density exemplifies a decreasing function.

A different type of ageing involves populations. A population ages not primarily by virtue of its people becoming older, but because fewer children are being born. Consequently a population can also become younger if its birth-rate rises. To give an example: the population of Russia is ageing as a result of both the individuals becoming older, and the birth-rate dropping. It has been estimated that the population of that country is going to drop by over twenty percent by the year of 2050, a fearful decline which, if not arrested, is likely to have far-reaching economic and demographic consequences.

Chapter Eighteen

Biomarkers

To return to biomarkers, some of those that show a progressive decline have been found to reveal an interesting, if controversial, feature, which is illustrated in Fig. 18.1 (see pages 32, 33). This shows the results of three studies, each of which illustrates how human biological functions show the simplest decline with age. The top one (Nette, Xi, Sun, Andrews & King, 1984) shows the fate of cells obtained from the skin and irradiated by ultraviolet radiation which damages the DNA they contain. The upper black circles record the degree to which recovery of the DNA can take place once irradiation has ceased. Note that, unavoidable scatter of the data notwithstanding, there is a marked, and statistically highly significant, decline with age: years are shown along the horizontal scale. In contrast, cells which have not been irradiated (empty circles) and therefore do not exhibit any recovery, hug the near-zero line independently of age.

The second graph (Schmidt & Peisch, 1986) refers to a pigment to be found at the back of the retina of the human eye. It is black (hence its name), and its concentration declines systematically with age irrespectively if one is dealing with blue, brown or hazel eyes, or those of indeterminate colour.

Lastly, the third example documents the age-related decline of human collagen III. This houses a gene, professionally referred to as COL3A1. Collagen is essentially a connective tissue, a sort of glue, found in the skin, the lung, blood vessels and also bone material. When boiled, it yields gelatin. In this instance, it was obtained from the lamina cribrosa, a tissue intertwining with the fibres leaving the eyeball on their path to the brain (Albon, Karwatowski, Avery, Easty & Duance, 1995).

You will have noticed that each graph contains a continuous straight line, and a broken continuation leading to the axis indicating age. This needs an explanation. The continuous lines are calculated from the observations, and represent the most probable course of the declines shown. Rectilinear changes in what is referred to as a variable are encountered in everyday life. Suppose we hail a taxi. The meter will show a starting price. Then the cost is going to increase with each mile or kilometre covered, if we ignore the ticking at traffic lights and other stops. In this instance, the straight-line graph will show a rise. A straight-line or linear graph is characterized by a constant rise or decrease, irrespective of the independent (given) variable, i.e. distance in our example.

However, it can be converted into a declining line when it is not the cost that is considered, but when the distance travelled is related to the amount of petrol in the tank. Let this be recorded at every completed mile or kilometre. This will give rise to a continuous line, as typified in Fig. 18.1. Now that we know the consumption in so much petrol per unit distance, we can determine by means of a broken line when the tank is going to run dry.

Thus the broken line in Fig. 18.1 does just that. In the top example it indicates the hypothetical age at which DNA will have become irreparable after ultraviolet irradiation. In the second example, the broken line indicates the age at which the retina will have lost all its black melanin. And the broken line in the third graph gives the age at which the skins used in giving rise to the data on collagen will become devoid of that substance.

Have you spotted that the hypothetical endpoints lie within some 15 years of one another? Their mean is 121 years. Now several analyses of much larger numbers of human biomarkers (see Weale, 1993) have shown that the most probable average zero age lies between 120 and 130 years.

It is important to justify the procedure since it is uncommon. The algebraic description of the continuous straight line is based on two constants, namely the intercept, the value they have at the zero

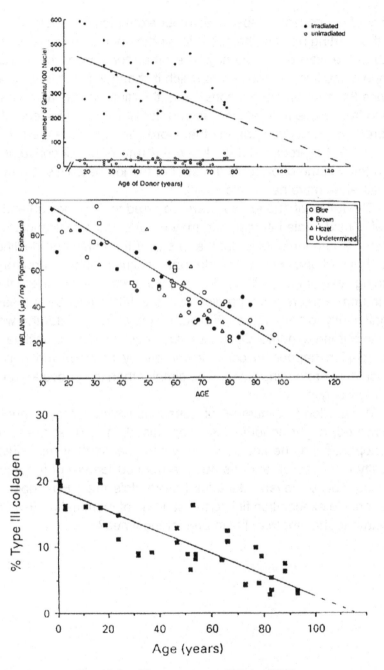

Fig. 18.1. Please see text (Weale, 1993).

point of the horizontal abscissa: in our examples, the meter reading at the starting point of the cab ride, or the volume in the tank of the cab at the start of the journey. The other constant is the so-called slope of the line, i.e. the rate at which it changes (rises or declines). Since the zero on the horizontal, e.g. the "final" point can be calculated from these two values, it is permissible to use it instead of the aforementioned intercept. In other words, the distance the cab can cover can be predicted if we know the rate of petrol consumption and the volume of petrol at the start of the journey. This is how the "final" ages have been calculated.

The range of 120 to 130 years (referred to on pages 33 and 86) is of considerable interest, and may be of some significance. If the above argument is valid, that is to say, if the combined decline of the biomarkers, which include both cellular ageing and organ ageing examples, relates to the life expectation, then those research workers who predict that the latter can be increased significantly to 150 or even 200 years, may have to alter the whole of the human genome. Of course, it is possible that the decline and expectation happen to coincide merely by chance, and that the problem of prolonging life may be simpler than the above argument has envisaged.

We are led to a number of questions which, if they cannot be answered, ought nonetheless to be raised. Why does the female menopause, i.e. the age at which women cease their reproductive faculty (~50 years), and, as such, a marked feature of the ageing process, occur several decades before life's end? How does our present life expectation fit into the scheme of mammals? How does it relate to the length of life in prehistoric times?

Chapter Nineteen

The Menopause

The problem with the menopause is that it has received a great deal of attention with particular emphasis on its possible evolutionary significance (see Shanley, Sear, Mace & Kirkwood, 2007). Why should it have evolved? It has been suggested that a late non-reproductive period in a woman's life will enable her to look after her youngest offspring. It is clear that, if she were able to reproduce until the end of her life, the development of the latest offspring might be compromised if not actually jeopardized. A parallel hypothesis postulates that the menopause provides for the existence of grand-mothers who can help in the upbringing of younger children.

This raises the question of how many offspring would have been brought into the world before contraceptive procedures had been thought of. A likely figure is 12, when we bear in mind that lactation inhibits conception. Thus, if a baby is born every two and a half years, a mother, who reached puberty by the time she was 14 or 15 years old, will be approaching her fortieth birthday when the likelihood of conception would be far on its path of decrease. This is illustrated in Fig. 19.1 which shows the decline of the formation of oocytes/follicles, i.e. female germ cells which develop into ova (de Bruin & te Velde, 2004). Note that the vertical scale is logarithmic, i.e. each unit decline corresponds to a reduction of a factor of ten. Incidentally, the number of oocytes in a woman's ovary is fixed; this provides a clue to the loss of fertility at some stage (cf. our simile of the car running out of petrol).

Shanley and her colleagues (2007) stress that both a terminal non-reproductive period on a woman's part and the grandmotherly help factor benefit the development of infants and young children

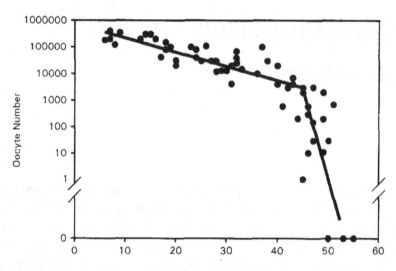

Fig. 19.1.   The loss of oocytes (vertical scale) against age (De Bruin & te Velde, 2004).

for the time required to enable them to stand on their own feet. We speculated above that any one woman would have a number of children so that a grandmother, obviously no longer in the first flush of youth herself, might have her hands full. On the other hand, the authors draw attention to the high rates of infant mortality even in historical times so that granny's load might be bearable. We shall return below to estimates relevant to prehistoric times.

The authors developed a statistical theory taking account of some of the above, and other, points, and tested it with data obtained for a population exhibiting "natural fertility", rare these days, but found in four villages in the Gambia. It transpired that the model satisfactorily agreed with a postulate that a grandmother played no role during the first year of a baby's life, but an increased one in subsequent years.

Other authors point out that the menopause protects against the development of still-births and birth defects, which tend to manifest with increasing maternal age (Pavard, Metcalf & Heyer, 2008). However, a crucial issue is whether the menopause is observable not only in non-human primates, but additionally in other animals.

Shanley and her colleagues' belief that the menopause is special to humanity is not accepted by all the experts. But others note that the menopause is widespread in the animal kingdom where helpful grandmothers might be in relatively short supply (see Wu, Zelinski, Ingram & Ottinger, 2005; Cohen, 2004). It is hard to conceive of a biological pressure that might lead to the menopause evolving in order to enable grandmothers to help in infant development. The principal pressure of evolution is procreation and, given the natural infant mortality rate in the distant past, one might be led to think that this was a more potent factor in letting some offspring survive successfully.

Chapter Twenty

Age in the Distant Past

Attention has been drawn earlier to the need of considering evolutionary problems in order to tear oneself away from the present, and to try and project the current situation into its past. Ageing is no exception, except that it is harder to obtain reliable facts.

In historic times, some information on the age of death could be obtained from tombstones. This was biased since tombstones cost money, and the poor part of the population — who almost certainly had a shorter life-expectancy than was true of the affluent minority — would not be represented. But it has been estimated that the life-expectancy, i.e. the age at which a cohort would have reached 50 percent survival, was about 20 years, give or take two or three years. It should be emphasised that life-expectancy was not invariable (anymore than it is today). There is some evidence to suggest that it actually decreased when farming became a way of life (Cohen, 1989).

Prehistoric times have no tombstones to send a message to us, and therefore we depend on the analysis of bones. It is understood that such information as may be obtained from their structure relates to their length of life, i.e. age, rather than the demographically entertained life-expectancy, i.e. one is dealing with individuals rather than populations. For example, an analysis of eighteenth-century bones from the crypt of Christ Church in London's Spitalfields suggests that the age of "young" bones has been overestimated, with the reverse of older bones (Colquhoun, 2008). By the way, sometimes one finds the term life-expectancy used in lieu of age or life expectation, liable to lead to misunderstandings; as one author puts it: "Life expectancy is mean age at death", a definition which clearly fails to take the concept of the cohort into consideration.

With the above provisos, it is generally understood that the lifespan was appreciably shorter prehistorically than in the recent centuries. Men, being hunters and gatherers, tended to be better fed than women — who were liable to suffer from malnutrition — and may hence have been less resistant to disease. The men's healthier lifestyle may have explained their higher expectancy by about three years than that of women, whose other main cause of death appears to have been childbirth (Jaffe, 1997).

There appear to have existed variations depending on geography. Thus in Hallein, Austria, female life-expectancy was 28 years, with the male value exceeding it by as much as seven years, whereas elsewhere both sexes might reach 35 to 40 years (Raimund, 1997). The difference is attributed to an especially hard life in the salt mines in Hallein together with poor hygienic conditions due to local restrictions. With reference to the previous section, it is not difficult to imagine that the menopause as we know it today may well have been a relatively rare phenomenon if most women died by the time they had reached the fifth decade of their lives. The occasional Celt woman appears to have reached an age of 69 years, and a man even 72 years. A similarly variable situation has been described for the Mississippi region (Wilson, 1997).

However, the general opinion is that life-expectancy at birth was in the high teens and twenties, with the upper thirties quoted as a maximum. For example, an Anglo-Saxon region shows a life-expectancy of 23 years. In Neolithic times, children in the Orkney Islands (off Scotland) outnumbered adults by three to one (Colquhoun, l.c.): this means that the age distribution must have been strongly pyramidal (when mortality at all ages is much the same its age distribution approaches a rectagonal shape). Life tended to terminate typically between 15 and 30 years of age. It has been estimated that a mere one point five percent of the local population was over 40, with very few reaching the age of 50. Today, the Orkneys do not enjoy the sunniest of climates; if this was also true in prehistoric times the caveat expressed above in connection with Hallein may apply also in this instance.

Chapter Twenty One

How Does Human Ageing Fit into the Animal Scheme?

Since all animal species have only limited lives, death and, with it, ageing, are a universal phenomenon (an exception appears to be a small jellyfish *Turritopsis nutricula* found in the Caribbean that can regenerate to a youthful form and be considered to be immortal). True, if life is cut short as a result of the individual experiencing violence due to a predator's voracity or as a result of an environmental event, marked signs of senescence may not appear as readily as they do when the body is left alone. In point of fact, protected individuals, be they human or animal, are liable to live longer than those exposed to the accidents of the wild. For example, the life expectation of rats is given as two to three years, but, looked after under laboratory conditions, they have been known to live for ten years or more.

Any pattern of inter-species ageing requires one or more biological variables characteristic of the species being compared with humanity. It is common knowledge that small animals live less long than large ones, not just because the former may be consumed by the latter: elephants easily outlive mice. The general point is illustrated in Fig. 21.1, which shows a plot of the maximum recorded age along the vertical against the body mass (expressed in grams) of 1701 creatures plotted against age on the horizontal (de Magalhães). The latter scale is somewhat unusual: the clue is provided by the underlying equation shown in the figure. It follows from it that log(age) = log5.58 + 0.146.log(weight). Now log(age) = 0

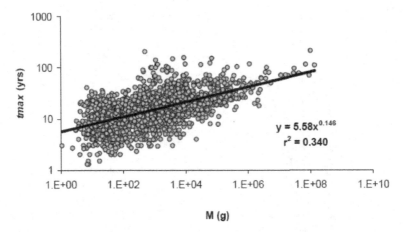

Fig. 21.1. Life-expectancy of different species against body mass (de Magalhães).

if log(weight) = −log5.58/0.146, and, when log(weight) = 0, log(age) = log5.58. Hence age is 5.58. However, since the graph uses logarithmic scales, we find the age corresponding to one year at 7.5 on the vertical scale, i.e. where the straight line hits it. This explains the use of the equation.

However, the scatter shows that, though statistically highly significant, this correlation is a bit of a bludgeon. It fails to explain any underlying reason for the observation; brain size might prove more explanatory. One of the earliest ground-breaking papers (Hart & Setlow, 1974) dealt with a mammalian repair process executed by fibroblasts; these are cells which combine collagen (see page 31) with extracellular tissue, and which play an important role in wound healing. Hart and Setlow irradiated their material with ultraviolet radiation, and then recorded the rate of recovery and its maximum extent (Fig. 21.2). Both variables were found to be approximately linear functions of the logarithm of the maximum lifespan. This is analogous to the example of the cost of a taxi ride, rising with the distance covered (see page 104). It will be appreciated that this is informative: a mechanism that acts rapidly, and has a greater potential of dealing with damage, is likely to help preserve its owner

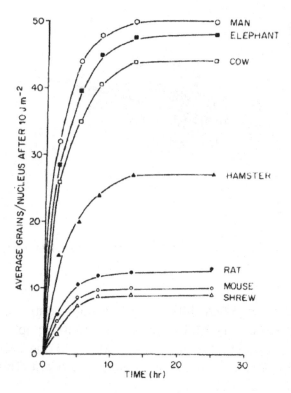

Fig. 21.2. Fibroblast recovery following irradiation for different species (Hart & Setlow, 1974).

for longer than if the reverse is the case. This then, is a basic example linking ageing with a biological entity.

The matter was extended in connection with the enzyme super-oxide dismutase (SOD), a scavenger of oxygen free radicals (see page 52), an element potentially noxious to biological tissues (Tolmasoff, Ono & Cutler, 1980). An enzyme facilitates a chemical reaction without itself taking part in it, much as is true of a catalyst.

The study covered primates mainly but not exclusively. The variables, other than the maximum lifespan, addressed by the authors were the bodyweight of young adult males of their chosen species, the weight-specific metabolic rate (SMR, see page 46), the maximal lifespan potential (MPL) calorie consumption (MCC), calculated by the equation MCC equals SMR times MLP, and, of course, measurements of SOD. The results are shown in Fig. 21.3, the key to the species numbers being provided in the attached table.

The authors stress that the values for the variables used should not be considered to be highly accurate because of the different sources they employed, yet they are sufficiently accurate to suggest that SOD equals a constant times MCC. The clear implication is that the high value of the human MCC, the maximal lifespan potential calorie consumption, rests on a biochemical basis and points to the evolution of our high longevity. It also seems to be clear that longer-lived species benefit from a high degree of protection from the ravages of free oxygen.

A greater variety of mammals was used in a study of a component of blood, namely mononuclear leukocytes, i.e. white blood corpuscles (Grube & Bürkle, 1992). It is well known that white blood corpuscles are protectors from wound infection: they appear as pus when a wound has been infected. A polymerase is an enzyme (all enzymes are characterized by the terminal -ase) the principal function of which is concerned with polymers of nucleic acids such as RNA and DNA. Its primary function is the polymerization, i.e. build-up (formation of polymers) of new DNA or RNA using an existing DNA or RNA template in the processes of replication and transcription of DNA.

The action of Poly(ADP-ribose) polymerase (n.b. PARP is a protein primarily involved in the repair of small areas of DNA damage or 'nicks') is strongly induced in the presence of DNA fragments, and acts in cellular repair and recovery when their DNA has been damaged (see page 104). The authors collected blood samples from several members of each species such as the rat, sheep, pig, horse, gorilla, elephant and man. Having "permeabilised" them, i.e. having made the cellular walls permeable to a variety of agents, they measured the maximal Poly(ADP-ribose) polymerase (PARP) activity as a function of maximal lifespan for the total of 13 mammalian species in all. It turned out that the PARP activity rose systematically — and statistically highly significantly — with the lifespan of the species in question. The

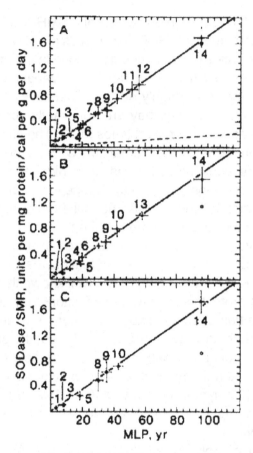

Fig. 21.3. Measurements of superoxide dismutase v. maximum lifespan potential. Data for the liver are shown at the top (A), for the brain in the centre (B), and for the heart at the bottom (C). The dotted line for the liver data represents the ratio when SMR was obtained from *in vivo* material from published data (Tolmasoff, Ono & Cutler, 1980).

The number key is as follows:

1. House mouse	2. Deer mouse	3. Common tree shrew
4. Squirrel monkey	5. Bush baby	6. Moustached tamarin
7. Lemur	8. African green monkey	9. Rhesus monkey
10. Olive baboon	11. Gorilla	12. Chimpanzee
13. Orangutan	14. Man	

PARP activity for the rat was 100 pmol whereas for man it was about five times as large. This serves to show that a small increase in PARP activity is very effective since the corresponding ratio of lifespans is 40 or 50 in favour of mankind.

It is clear that we are facing a situation closely related to that illustrated by Fig. 21.3: given a proven protective biological compound, the higher its quantity per body mass the more likely is the species going to exhibit a longer lifespan. This view needs to be taken with a pinch of salt: it might lead to the conclusion that obesity is the gateway to a long life (see pages 67, 72, 80, 85). As against that, body-height rather than mass tends to reduce mortality from stroke. During the last century there has been a well-documented tendency for each generation to be taller than its predecessor, perhaps as a result of a better understanding of dietary influences, and the role of lifestyle. The theoretically expected rise in life-expectancy (which is happening, but it is not established as being due to the aforementioned factors) may well be counteracted by the increasing prevalence of obesity and indifferent lifestyles. Thus the above examples have to be seen as informative when the donors, human and animal, can be considered to be free from any known pathology.

By the way, the above studies might lead us to conclude that longer-living species might be blessed with longer telomeres (see Fig. 12.2), and that is why they live longer. Not so. It has been found that mice have longer telomeres than we have; there is, however, no evidence that mice outlive the longer-livers.

Chapter Twenty Two

From End to Start

It has been suggested during recent times not only that there exist specific seasons of birth but also that this may affect your age. The latter point will be shown later in this section to follow logically from the former.

That there may appear a seasonal variation in the statistics of birth — or more probably of conception — is not altogether surprising. In India, to quote a striking example, there is a relative dearth of births in some states in the spring: the hot summer nights of the season of the Monsoon offer a powerful deterrent to indulging in hot embraces (cf. Prasad, Srivastava, Bhushan & Jain, 1969; Bernard, Bhatt, Potts & Rao, 1978). The latter authors point at the possibility that a high scrotal temperature may be counter-productive to a successful insemination during very warm months.

In the African Gambia, where there is a "hungry season", young adults are likely to die prematurely as compared with those who have succeeded to avoid being born when maternal nutrition was inadequate (Moore, Cole, Poskitt, Sonko, Whitehead, McGregor & Prentice, 1997). The "hungry" season is likely to have included the period of the foetal development with adverse long-term effects on the efficiency of the immune system. We have to understand that seasonality is not necessarily related to length of life; it merely records when, in the course of a year, some events may have occurred.

Environmental influences on the statistics of birth are not confined to tropical or sub-tropical regions (Miura, 1987). The annual distribution associated with Europe (Fig. 22.1) is double-humped, the major peak in late winter/early spring being attributed to the

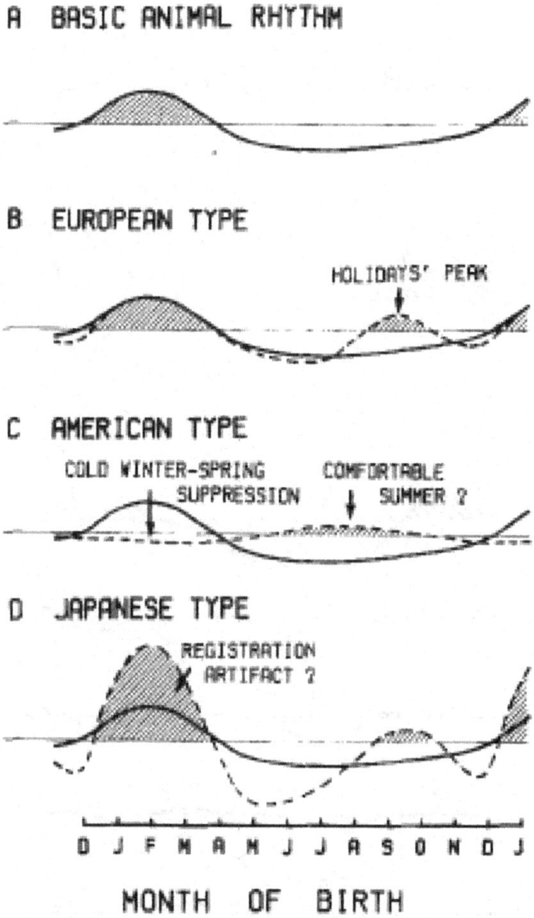

A BASIC ANIMAL RHYTHM

B EUROPEAN TYPE

HOLIDAYS' PEAK

C AMERICAN TYPE

COLD WINTER-SPRING   COMFORTABLE
SUPPRESSION   SUMMER ?

D JAPANESE TYPE

REGISTRATION
ARTIFACT ?

D J F M A M J J A S O N D J

MONTH OF BIRTH

Fig. 22.1. Different types of seasonality (Miura, 1987; courtesy Birkhäuser Verlag).

clement nights in late spring/early summer, while the minor hump has been attributed to frolicking during and after the Festive season. The simpler American picture might be due to a difference in domestic heating provisions. Incidentally, it would be a mistake to assume that birthrate distributions are invariable. Miura points out that they have changed systematically in Japan during the course of the previous century (Fig. 22.2), but he does not believe "that the recent changes of birth seasonality or the disappearance

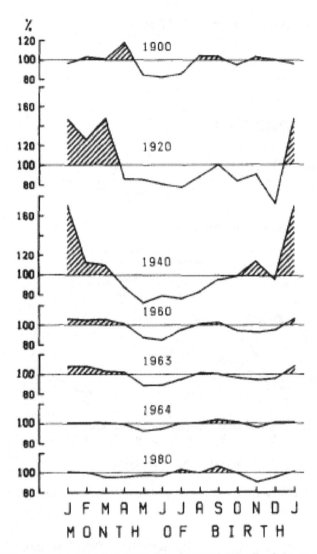

Fig. 22.2. The secular flattening ("deseasonalization") of the annual pattern in Japan (Miura, 1987; courtesy Birkhäuser Verlag).

of the prominent spring peak of birth in Japan can be caused simply by the recent so-called deseasonalization [*sic*] caused by the changes in, or improvements of, accommodation and sanitary conditions".

It is also the case that the cuddling factor associated with human comfort is not the only one to show an influence on the reproduction statistics. There is evidence that, in addition to nutritional seasonality, there are seasonal effects relating to infectious diseases which can have a long-term effect on age (Muñoz-Tuduri & Garcia-Moro, 2008).

Thus a study carried out in the Spanish island of Minorca showed that the seasonal distribution of death of children less than one year old and also of those younger than 15 years of age mimicked the seasonal variation of birth. After the age of 15, the number of deaths during the summer was reduced, a result hypothetically attributed to a lower susceptibility to degenerative diseases in those born in the summer, and presumably also to the adequate nutrition available in Minorca in the autumn.

Muñoz-Tuduri & Garcia-Moro reported seasonal variations for births, infant deaths, neonatal deaths and postnatal deaths, similar to those shown for the two European countries in Fig. 22.3.

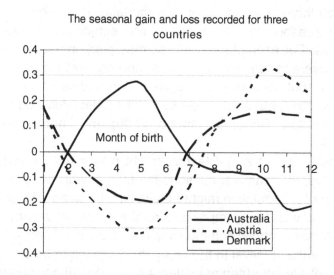

Fig. 22.3. The seasonal gain (vertical scale: positive) and loss recorded for Australia, Denmark and Austria (smoothed data after Doblhammer & Vaupel, 2001). Note that the seasonal gain or loss amounts to a mere fraction of a year. For details please see text.

Their pattern is to be found in other parts of Europe, particularly in the Mediterranean region where the seasons do not show appreciable geographic variations at any one longitude. The peak mortality in Minorca in the summer months is attributed to gastro-intestinal problems, partly linked to a poor drinking-water quality.

A long-term influence exercised by the season of birth on those aged 50 years and over is shown in Fig. 22.3 (simplified, after Doblhammer & Vaupel, 2001). The two approximately parallel sets of data relate to Denmark and Austria (the lower line), whereas the mirror image describes the position in Australia. Remember that it is winter down under when Europe enjoys (?) the summer months. The authors tested their results proposing possible explanatory hypotheses.

(a) It was assumed that the prime mover was the interaction between the season of mortality and age. The example given is that people born in April are older than those in November when there is a season of high mortality. It may be mentioned that seasons of high mortality are subject to climatic influences. For example, mortalities may peak in the winter when the average temperature is low, and domestic heating inadequate; but they may also peak in the middle of the summer when the temperature can be fatal as decades ago in Spain (see below: Reher & Gimeno, 2006) and, more recently, in Paris.

(b) Possible latent social factors might play a role. These might involve, for example, religious matters, in which case different parts of a population might be similarly involved. The period of Lent may be found impressed on the seasonal pattern notably in Roman catholic communities (see Reher & Gimeno, l.c.).

(c) Differential survival in the first year of life might be operative.

(d) Potentially intra-uterine influences may be definite determining factors. Like all such studies, the research workers depend on records, which must be accurate in order to be of any value. It is therefore the case that only relatively advantaged societies will be able to afford the necessary means. There is, however, a notable, and very useful, exception: records kept in the Gambia have enabled research to be extended over several generations (see Moore, Cole, Poskitt *et al.*, l.c.).

Doblhammer and Vaupel report, as do other authors, that foetal growth as deduced from birth-weight is an indicator of future disease (see also for example Barker, Bull, Osmond & Simonds, 1990). They show that individuals born between April and July, i.e. after the winter months, were suffering from reduced intra-uterine growth. Since birth-weight shows a seasonal variation, this is likely to be reflected in the seasonal-disease pattern; if, and not only if, the immune system is involved, an effect of the life-expectation may follow.

The authors conclude by reminding us that a quarter of the variation in the human life-expectation is determined genetically. For example, it has been said that the maternal age at death plays a role but only up to the age of 60 years. Another quarter of the variation depends on the early part of life, and the influences the individual is subjected to, as much of the remaining half is subject to adult influences, i.e. lifestyle, and environmental effects. The authors hold that though the influence of the season of birth on the variation in the length of age is small (though statistically significant), as follows from the vertical scale of Fig. 22.3, it is important from the point of view of public-health policy if not for any other reason than that any clinically useful forecast is valuable.

Several authors (e.g. Miura; Reher & Gimeno) have shown that there are progressive changes in the seasonality of births, and the dates of conception deduced from the dates of birth; in general the variation due to seasonality tends to decline on a secular scale. This may be due to an increasing human control of the environment, to the use of contraceptive methods, or to hitherto unestablished causes. Thus, there exist profound differences as between (Spanish) rural and urban communities, the latter showing paradoxically the more marked seasonality.

Although Reher and Gimeno's study is concerned mainly with the effect of the date of birth on infant and childhood mortality they stress, as we have seen above, that the date exerts a long-term influence on the individual.

This transpires also from a couple of recent studies (Costa & Lahey, 2006; Doblhammer, 2002). The data studied by Costa & Lahey related to white soldiers in the American Union army fighting in the Civil War (1861–65), and alive in 1900. Seasonality showed spring and summer to have been bad annual quarters to have been born in. The authors were able to exclude social and educational factors, effects of the Civil War and also mortality effects in early life. Typically, former members of the Union Army born in the spring and summer were found more likely to die of cardiac problems and stroke than those born at other times of the year. The odds relating to these and other conditions ranged from 1.3 to 1.7, and thus were not negligible.

Doblhammer published one of the most extensive studies in the field (Doblhammer, 2002, l.c.). She studied 15 million death certificates covering the period of 1989 to 1997. Those born in the autumn live five months and eight days (0.44 years) longer than those born in the spring. Many parameters were examined; they included the individual's ethnic group, region of birth (urban or rural), marital status and degree of education and, in addition, gender (see also: Lerchl, 2004).

It was also found that the seasonal variation of birth was related to the major causes of death, such as cardiovascular diseases, cancer (especially lung cancer, which is surprising as the degree of smoking plays a considerable role in this condition) and also infectious diseases, which probably implicated the state of the immune system. Some samples of the variable parameters are shown in Fig. 22.4.

Note that some of the data show much more systematic variations than others: respiratory diseases, infectious diseases, conditions associated with the digestive system, pneumonia and influenza stand out in that respect. As might be expected,

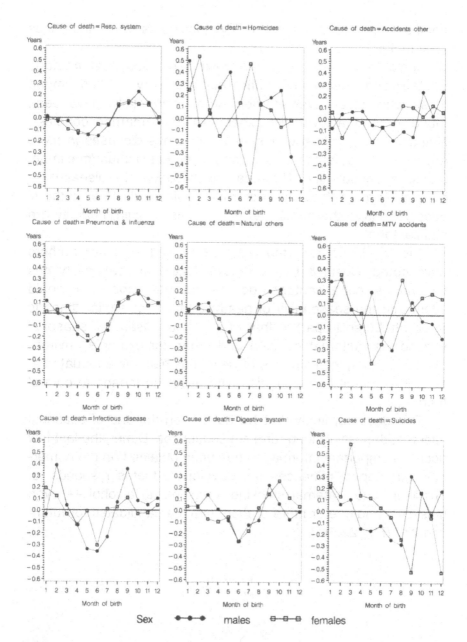

Fig. 22.4. Samples of variable parameters in seasonality (Doblhammer, 2002).

deaths from motor-vehicle (MTV) accidents are much less systematic.

The author concludes from her extensive study that marriage is stabilizing: the month-of-birth effect is very marked among those who have never married (who, incidentally, have been found to have a lower life-expectancy than married couples). Nutrition and potential exposure to infectious diseases *in utero* are held to be potent factors in causing large minimum/maximum ratios in seasonality. The seasonality linked to diseases is caused by more than one factor in keeping with other authors' observations to the effect that the causes of child mortality are multifactorial.

Just as Miura reported (Fig. 22.2), and as other authors have noted, Costa & Lahey (2006) state that seasonality has flattened since a century ago. As indicated above, it would seem that the seasonality of death reported for earlier times is to be attributed in part to that of infectious diseases, seasonal variation in maternal nutrition and other, for example environmental, factors, the influence of which has become attenuated at least in privileged communities who have the means to control them.

This does not imply that we have seen the end of seasonal effects with possibly long-term consequences. During the last few decades, important information has been obtained on the numerous extraneous influences the developing foetus responds to. These include maternal smoking, drug intake, alcohol — even music —, and one or more of them may be candidates for future concern (Fig. 22.5).

Doctor examining an expectant woman: "Mrs.Smith, this text message seems to be addressed to YOU!"

Mrs Smith: "But I don't possess a mobile."

Doctor: "It reads: Mummy, make sure you deliver me on the best day of the year, pleeaase."

Fig. 22.5. "At the beginning was the Word ..."

References

Albon J., Karwatowski W.S.S., Avery N., Easty D. & Duance V.C. (1995). Changes in the collagenous matrix of the aging human lamina cribrosa, *British Journal of Ophthalmology*, 79:368–75.

Barker D.J., Bull A.R., Osmond C. & Simmonds S.J. (1990). Fetal and placental size and risk of hypertension in adult life, *British Medical Journal*, 301:259–62.

Bergman J. "Darwin's Teaching of Women's Inferiority", Institute for Creation Research, 1994, http://www.icr.org/article/378/.

Bernard R.P., Bhatt R.V., Potts D.M. & Rao A.P. (1978). Seasonality of birth in India, *Journal of Biosocial Science*, 10:409–21.

Cohen A.A. (2004). Female post-reproductive lifespan: A general mammalian trait, *Biological Review,* 79:733–50.

Cohen M.N. (1989). *Health and the Rise of Civilization,* pp. 131–41, Yale University Press, New Haven & London.

Colquhoun I.A. (2008). Life expectancy in prehistory, http://archaeology.suite101com/article.cfm/prehistoric_population.

Comfort A. (1962). *The Biology of Senescence* (3rd edition), Churchill Livingstone, London.

Costa D.L. & Lahey J.N. (2006). Predicting older age mortality trends, *Journal of the European Economic Association*, 3:487–93.

Darwin C. (Nora Barlow, Ed.). (1958). *The Autobiography of Charles Darwin, 1809–1882*, pp. 232–3. W. W. Norton & Co., Inc, New York.

de Bruin J.P. & te Velde E.R. (2004). Female reproductive ageing: Concepts and consequences. In: *Preservation of Fertility,* eds. Tulandi T. & Gosden R.G. Taylor & Francis, London.

de Magalhães J.P. *Comparative Biology of Ageing.* www.senescence.info/comparative.html.

Doblhammer G. (2002). Differences in Lifespan by Month of Birth for the United States: The impact of early life events and conditions on late life mortality, *Max Planck Institute for Demographic Research Working Paper 2002-29.*

Doblhammer G. & Vaupel J.W. (2001). Lifespan depends on month of birth, *Proceedings of the National Academy of Sciences of the United States of America,* 98:2934-9.

Engeland C.G., Bosch J.A., Cacioppo J.T. & Marucha P.T. (2006). Mucosal wound healing: The roles of age and sex. *Archives of Surgery* 141:1193-7.

Grube K. & Bürkle A. (1992). Poly[ADP-ribose] polymerase activity in mononuclear leukocytes of 132 individuals from 13 mammalian species correlates with species-specific life span, *Proceedings of the National Academy of Sciences of the United States of America,* 89:11759-63.

Hart R.W. & Setlow R.B. (1974). Correlation between deoxyribonucleic acid [DNA] excision-repair and life-span in a number of mammalian species, *Proceedings of the National Academy of Sciences of the United States of America,* 71:2169-74.

Hollingsworth J.W., Hashizume A. & Jablon S. (1965). Correlations between tests of aging in Hiroshima-subjects — An attempt to define "physiologic age", *Yale Journal of Medicine,* 38:11-26.

Jaffe A.J. (1997). Life expectancy among prehistoric North Americans, www.popline.org/docs/0561/014715.html; R. Karl: Celtic-L — The Celtic culture list.

Lerchl A. (2004). Month of birth and life expectancy: Role of gender and age in a comparative approach, *Naturwissenschaften,* 91:422-5.

Miura T. (1987). The influence of seasonal atmospheric factors on human reproduction, *Experientia,* 43:48-54.

Moore S.E., Cole T.J., Poskitt E.M.E., Sonko B.J., Whitehead R.G. & McGregor I.A. (1997). Prentice M.A.: Season of birth predicts mortality in rural Gambia, *Nature,* 388:434.

Muñoz-Tuduri M. & Garcia-Moro C. (2008). Season of birth affects short- and long-term survival, *American Journal of Physical Anthropology,* 135:462-8.

Nette E.G., Xi Y.-P., Sun Y.-K., Andrews A.D. & King D.W. (1984). A correlation between aging and DNA repair in human epidermal cells, *Mechanisms of Aging & Development*, 24:283–92.

Pavard S., Metcalf E. & Heyer E. (2008). Senescence of reproduction may explain adaptive menopause in humans: A test of the "mother" hypothesis, *American Journal of Physical Anthropology*, 136:194–203.

Prasad B.G., Srivastava R.N., Bhushan V. & Jain V.C. (1969). A study of the seasonal variation of births at maternity hospitals of some medical colleges in India, *Indian Journal of Medical Research*, 57:718–29.

Raimund K. (1997). CELTIC-L — The Celtic Culture List.

Reher D.S. & Gimeno A.S. (2006). Marked from the outset: Season of birth and health during early life in Spain during the demographic transition, *Continuity and Change*, 21:107–29.

Russell N. (1988). Oswald avery and the origin of molecular biology, *British Journal of the History of Science*, 21:393–400.

Schmidt S.Y. & Peisch R.D. (1986). Melanin concentration in normal human retinal epithelium, *Investigative Ophthalmology and Visual Science*, 27:1063–7.

Shanley D.P., Sear R., Mace R. & Kirkwood T.B.I. (2007). Testing evolutionary theories of menopause, *Proceedings of the Royal Society B*, 274:2943–9.

Strehler B.L. (1962). *Time, Cells, and Aging,* Academic Press, New York.

Tolmasoff J.M., Ono T. & Cutler R.G. (1980). Superoxide dismutase: Correlation with life-span and specific metabolic rate in primate species, *Proceedings of the National Academy of Sciences of the United States of America*, 77:2777–81.

Weale R.A. (1993). Have human biological functions evolved in support of a lifespan? *Mechanisms of Aging & Development*, 69:65–77.

Wilson D. (1997). Average age of death, and control over reproduction, contraception, and childbearing, p. 125. In: *Women in Prehistory* (eds. Claassen C. & Joyce R.A.), University of Pennsylvania Press.

Wu J.M., Zelinski M.B., Ingram D.K. & Ottinger M.A. (2005). Ovarian aging and menopause: Current theories, hypotheses, and research models, *Experimental Biology and Medicine*, 218:818–28.

Index

Accommodation, 37, 38
Age-related disease, 16
Age-related maculopathy, 40
Alcohol, 17, 66, 72–75, 80, 85,
    91, 92, 94, 95, 126
Alzheimer's disease, 59, 60–63,
    93
Ancestors, 16
Ancient Rome, 15
Antioxidant, 40, 52, 76, 95
Apoptosis, 56
Arteriosclerosis, 32
Arthritis, 66
Atherosclerosis, 79

Babies, 38
Backpain, 65
Basal metabolic rate (BMR), 46,
    49
Biological age, 24–26
Biological clock, 23, 24
Biomarker, 32–34, 102–104,
    106
Birth-rate, 102
Blind-spot, 36
Blood, 51, 56, 74, 78, 82, 115
Blood corpuscles, 70, 115

Blood-pressure, 28, 48, 78–80,
    83, 102, 103
Blood vessels, 27, 28, 32, 50,
    102, 103
Body temperature, 46
Bone, 30, 31, 56, 63–65, 67, 69,
    70, 102, 103, 110
Boxing, 62
Brain, 32, 33, 36, 46, 56, 57, 60,
    63, 70, 71, 81–83, 104, 113,
    116
Breast cancer, 58, 71, 72, 76, 94

Calcium, 31, 65, 66, 76
Caloric restriction, 45, 46, 51
Cancer, 16, 22, 56, 68–70, 72, 74,
    124
Cancer of the prostate *see*
    Prostate cancer
Carbohydrate, 46
Cardiac, 124
Cardiovascular condition, 61
Cardiovascular diseases, 124
Cartilage, 30, 66, 67
Cataract, 91–93
Cell, 32, 33, 49, 54–57, 70, 71,
    101–103, 106, 113, 115

Cholesterol, 79, 80, 83
Chromosomal, 102
Chromosome, 53, 55
Climate, 37, 122
Collagen, 27, 31, 103, 104, 113
Colour, 38–40
Cone, 36, 39
Cornea, 36, 45, 51

Dehydration, 48
Diabetes, 16, 74, 80, 83
Diabetics, 81
Diet, 32, 33, 39, 45, 62, 66, 68,
    69, 71, 74, 76, 77, 79, 80,
    83, 90, 94–97, 117
Diet restriction, 44, 52, 94
Dilator, 42
Disposable soma, 50
DNA, 28, 50, 52, 55, 56, 103,
    104, 115
Drink, 11, 62, 72, 91, 92
Drug, 17, 85, 126

Elasticity, 21, 27
Elastin, 21, 27
Environment, 17, 20, 39, 49, 50,
    71, 112, 118, 123, 126
Evolution, 17, 19, 20, 34, 49, 50,
    101, 107, 109, 110
Eye, 35, 41, 45, 51, 52, 56, 58,
    70, 91, 102–104,

Fall, 81, 86
Fashions, 9
Fat, 85, 95
Fibroblasts, 113

Fish oil, 68
Food, 44, 49, 85
Fracture, 31, 81
Free oxygen, 51, 52
Free radicals, 57
Frequency, 7, 8

Gender, 33, 34, 61, 65, 93, 124
Gene, 17, 51, 59, 71, 76, 77, 79,
    98, 103, 123
Genetic, 18, 19, 21, 28, 33, 34,
    55, 57, 58, 62, 66, 67, 70, 74,
    80, 83, 101
Genotype, 19
Germ, 56
Germ cell, 50, 56, 107
Gerontology, 101
Grip, 87
Grip strength, 26, 32, 102

Hair, 50
Healing, 27
Hearing, 7, 8, 35, 58
Heart, 46, 51, 59, 78, 80, 81,
    116
Heart disease, 57, 78, 83, 94
Hormonal, 64, 68
Hormone, 76
Hypothermia, 28

Immune system, 25, 57, 67, 71,
    86, 118, 123, 124
Immunity, 45
Infant death, 16

Kidney, 66, 80

Lactation, 107
Language, 5
Lens, 36–39, 41, 43, 91, 92
Life-expectancy, 15, 16, 32, 68, 80, 94, 95, 110, 111, 117, 126
Life expectation, 16, 106, 110, 112, 123
Lifespan, 17, 50, 57, 66, 111, 113–117
Lifestyle, 10, 64, 90, 94, 117, 123
Lipofuscin, 52
Liver, 56, 116
Longevity, 16, 33, 44, 51

Macular degeneration, 40
Macular pigment, 39, 40
Maculopathy, 92, 93
Mediterranean diet, 62
Melanin, 22, 104
Melanoma, 22
Menopause, 31, 62, 71, 72, 80, 106–109, 111
Mice, 57
Muscle, 26, 28–30, 32–34, 42, 46, 49, 50, 52, 57, 66
Muscular, 57, 78, 80
Music, 58, 84, 126

Neoplasm, 45
Nitrogen, 55
Nucleotide, 55
Nucleus, 54

Obese, 31, 80
Obesity, 67, 72, 74, 85, 93, 117

Oestrogen, 71, 72
Oocyte, 107
Osteoarthritis, 67, 68
Osteoporosis, 31, 64, 66
Oxygen, 57, 114, 115

Phenotype, 19
Phosphorus, 55
Pneumonia, 16, 66, 81, 86, 124
Prostate cancer, 58, 64, 69, 73, 75, 76, 94
Protein, 46, 52, 53, 55, 62
Puberty, 16, 107
Pupil, 36, 41–43

Queen's (or King's) English, 5

Radioactivity, 75
Rat, 45, 96, 97, 112, 117
Reaction time, 49
Reading glasses, 14, 36, 37
Received English, 5, 6
Rejuvenation, 12
Religion, 93, 94, 122
Repair, 33, 50, 51, 53, 56
Reproduce, 49
Reproduction, 34, 101, 106, 107, 121
Retina, 36, 37, 39, 40, 41, 43, 52, 56, 91, 103, 104
Risk-factors, 68
Rods, 36, 39
Rugby, 62, 67

Salt, 48, 79, 80
Season, 118, 123, 126

Seasonality, 118, 119, 121,
123–126
Skeleton, 63
Skin, 14, 20–23, 27, 28, 32, 50,
103, 104
Smoke, 94
Smokers, 33, 62, 124, 126
Smoking, 17, 23, 24, 40, 52, 66,
68, 74–76, 79, 80, 85, 91–93,
96, 124
Somatic cells, 33, 50, 56
Sound, 7
Species, 50, 112–115, 117
Sphincter, 42
Spinal, 65, 67, 70
Spine, 29
Statin, 62
Stem cells, 51
Stereotype, 3
Stomach, 69
Stress, 17, 47
Stroke, 51, 61, 78, 82, 83, 85,
117, 124

Sugar, 55
Sunlight, 21, 22
Superoxide dismutase, 52, 114,
116
Surgery, 13

Teeth, 52
Television screen, 11
Telomerase, 55–57
Telomere, 55–57, 102, 117
Temperature, 37, 64, 65, 76
Testosterone, 64, 65, 76
Thyroid, 66
Tombstone, 110
Transmissivity, 38
Transparency, 91
Tuberculosis, 16

Vertebrae, 30, 65–67
Vital capacity, 32
Vitamin, 46, 66, 77

Wear and tear, 17, 21, 30, 52, 65